IX International Conference on Modern Techniques of Plasma Diagnostics and Their Application 2014

Journal of Physics: Conference Series Volume 666

Moscow, Russia
5 – 7 November 2014

Editors:

Andrey Kukushkin
Alexander Kukushkin
Elena Baronova
Emil Dodulad

ISBN: 978-1-5108-1958-0
ISSN: 1742-6588

Printed from e-media with permission by:

Curran Associates, Inc.
57 Morehouse Lane
Red Hook, NY 12571

Some format issues inherent in the e-media version may also appear in this print version.

Printed by Curran Associates, Inc. (2016)

For permission requests, please contact the Institute of Physics
at the address below.

Institute of Physics
Dirac House, Temple Back
Bristol BS1 6BE UK

Phone: 44 1 17 929 7481
Fax: 44 1 17 920 0979

techtracking@iop.org

Additional copies of this publication are available from:

Curran Associates, Inc.
57 Morehouse Lane
Red Hook, NY 12571 USA
Phone: 845-758-0400
Fax: 845-758-2634
Email: curran@proceedings.com
Web: www.proceedings.com

Table of contents

Volume 666

IX International Conference on Modern Techniques of Plasma Diagnostics and their Application
5–7 November 2014, Moscow, Russia

Accepted papers received: 16 November 2015
Published online: 11 January 2016

Preface

Papers

Journal of Physics: Conference Series **666** (2016) 011001 doi:10.1088/1742-6596/666/1/011001

IX International Conference on Modern Techniques of Plasma Diagnostics and their Application

A S Savjolov and E I Dodulad
National Research Nuclear University MEPhI (Moscow Engineering Physics Institute), Kashirskoe shosse 31, Moscow, 115409 Russian Federation

E-mail: savjolov@plasma.mephi.ru, EIDodulad@mephi.ru

The IX Conference on "Modern Techniques of Plasma Diagnosis and their Application" was held on 5 – 7 November, 2014 at National Research Nuclear University MEPhI (NRNU MEPhI).

The goal of the conference was an exchange of information on both high-temperature and low-temperature plasma diagnostics as well as deliberation and analysis of various diagnostic techniques and their applicability in science, industry, ecology, medicine and other fields. The Conference also provided young scientists from scientific centres and universities engaged in plasma diagnostics with an opportunity to attend the lectures given by the leading specialists in this field as well as present their own results and findings.

The first workshop titled "Modern problems of plasma diagnostics and their application for control of chemicals and the environment" took place at Moscow Engineering and Physics Institute (MEPhI) in June 1998 with the support of the Section on Diagnostics of the Council of Russian Academic of Science on Plasma Physics and since then these forums have been held at MEPhI every two years. In 2008 the workshop was assigned a conference status.

More than 150 specialists on plasma diagnostics and students took part in the last conference. They represented leading Russian scientific centres (such as Troitsk Institute of Innovative and Thermonuclear Research, National Research Centre "Kurchatov Institute", Russian Federal Nuclear Centre – All-Russian Scientific Research Institute of Experimental Physics and others) and universities from Belarus, Ukraine, Germany, USA, Belgium and Sweden. About 30 reports were made by young researchers, students and post-graduate students.

All presentations during the conference were broadcasted online over the internet with viewers in Moscow, Prague, St. Petersburgh and other cities.

The Conference was held within the framework of the Centre of Plasma, Laser Research and Technology supported by MEPhI Academic Excellence Project (Russian Ministry of Education and Science contract 02.a03.21.0005 of August 27th 2013).

Papers selected by the Program Committee for publishing were reviewed under control of invited editors Prof. Andrey Kukushkin, Dr. Sci. Alexander Kukushkin, Dr. Sci. Elena Baronova, Dr. Emil Dodulad.

We would like to thank heartily all of the speakers, participants and organizing committee members for their contribution to the conference.

Journal of Physics: Conference Series **666** (2016) 011002 doi:10.1088/1742-6596/666/1/011002

Peer review statement

All papers published in this volume of *Journal of Physics: Conference Series* have been peer reviewed through processes administered by the proceedings Editors. Reviews were conducted by expert referees to the professional and scientific standards expected of a proceedings journal published by IOP Publishing.

Journal of Physics: Conference Series **666** (2016) 012001 doi:10.1088/1742-6596/666/1/012001

Advanced probe edge diagnostics for fusion devices

G Van Oost[1]
Department of Applied Physics, Ghent University, St.Pietersnieuwstraat 41, B-9000 Gent, Belgium
National Research Nuclear University MEPhI (MEPhI), Kashirskoe shosse 31, Moscow, 115409 Russian Federation

E-mail: guido.vanoost@ugent.be

Abstract. Summary paper on advanced probe edge diagnostics techniques for fusion devices is presented.

1. Introduction

Since the 1990s it became increasingly clear that boundary plasmas play a major role in magnetic fusion devices (MCD), and strongly relate to and even dominate central plasma processes. On the one hand, the conditions of the boundary plasma are crucial to obtain high fusion triple products; on the other hand, plasma-surface interactions, a sufficiently low impurity concentration in the fusion volume, heat removal and helium exhaust which directly relate to the boundary plasma, have emerged as equally important goals, and even more difficult to reach in the state of self-sustained thermonuclear burn. Successful resolution of these issues is critical to establish the viability of the MCD concept for a fusion power reactor.

All these requirements invoke a complex interplay of core plasma, boundary plasma, and atomic and surface physics. Hence, there is an ongoing effort regarding plasma diagnostics which is essential to improve our understanding of the MCD boundary. The employment of edge diagnostics has to take into account the specific properties of the plasma edge, taken here to be synonymous with the plasma boundary. In magnetic confinement devices the plasma is confined within closed magnetic surfaces, normally generated by a combination of fields due to external conductors and/or by currents flowing in the plasma. A thorough understanding of the processes occurring in the plasma boundary is necessary for the optimization and confident up-scaling of today's results to future plasma machines and finally to a fusion reactor.

The plasma boundary can be divided into two regions: the region inside the separatrix or the Last Closed Flux Surface (LCFS), and the scrape-off layer (SOL) which is outside the LCFS. All experiments prior to the 1970's were not large enough to produce core plasmas well separated from the influence of the walls; strictly speaking, up to that time only boundary plasmas have been investigated. Energy and particles are transported from the plasma core along steep gradients (scale lengths of order 10 mm) of plasma parameters characterizing the first region, the radiating layer, which is also characterized by the significant presence of neutral particles. Ions generating from the plasma core are lost to the surrounding wall structures and neutralized with reemission (recycling

[1] To whom any correspondence should be addressed.

loop) of hydrogenic species and release of wall impurities by physical and chemical mechanisms. The species and properties (density, velocity, spatial distribution, radiation, etc) of the neutral particles in the radiating layer are determined by plasma-wall interaction and transport processes. The radial extent of this radiating layer inside the LCFS is determined by the ionization lengths of the neutral particles, and is of order 10 cm. The SOL refers to the region outside the LCFS where magnetic field lines are open, i.e. they intersect material surfaces like limiters or divertor plates; hence, the SOL is characterized by the competition between transport parallel and perpendicular to the magnetic field, and is directly related to the requirements for power and particle exhaust.

In contrast to the core plasma of a tokamak, axisymmetry does not exist for the boundary plasma which has a complex structure determined by the magnetic field topology, the very non-uniform walls, and the transport properties of the plasma along and across the magnetic field. In many cases the boundary structure is fully three-dimensional and various asymmetries can occur. Furthermore, the boundary plasma can be subject to rotations, auxiliary heating, pellet injection, magnetic field ergodization, etc. The experimental characterization of such a complex and turbulent edge plasma-where more quantities (such as neutral impurity and hydrogen influx ratio, flow speeds of hydrogen and impurity ions, heat- and particle outflux rates to surfaces, etc.) need to be known than in the central plasma imposes strong requirements to theoretical models as well as to diagnostics. Furthermore, due to the sometimes limited reproducibility of specific discharge conditions, it is necessary to measure simultaneously the (radial profile of) plasma parameters (e.g. using fast probe manipulators; see Section 2).

2. Plasma boundary diagnostics by probes

The understanding of the plasma edge and the control of edge conditions critically depend on measurements of the local plasma parameters. Reviews of techniques used for edge plasma diagnosis can be found in [1, 2]. Probe measurements (one of the earliest approaches in plasma diagnostics, dating back to I. Langmuir) complement spectroscopy by providing detailed profiles of local plasma parameters, as well as quantities like electric fields in the edge plasma which are difficult to determine spectroscopically, and the role of which in plasma confinement and exhaust is now widely recognized [3]. Furthermore, electrostatic turbulence-driven transport, which is generally believed to be the origin of anomalous edge particle transport in tokamaks, can only be fully evaluated with probes [4, 5]. In plasmas in which probes can survive, this diagnostic remains the easiest and most accurate way to measure local particle fluxes. This means that frequently only the edge is accessible, but its importance fully justifies the continued use of probes. It is wrong to assume that because of plasma-probe interaction probes are of less value than other diagnostic techniques: the presence of (much larger) surfaces in the plasma boundary as limiters, walls or divertor plates is inevitable.

Tokamak plasma diagnosis by electrical probes has been reviewed by Matthews [6]. Probe techniques for plasma edge diagnostics in magnetic confinement devices can be broadly divided into two categories: electrical (active) and surface collection (passive) methods. With electrical probes real time measurements are obtained as electrical currents drawn from the plasma; they encompass different types of Langmuir probes, and advanced electrical probes [6] such as gridded energy analyzers, mass spectrometers and probe arrays for the study of turbulence. With collector probes (not treated in this paper), on the other hand, the deposition of impurities on a collecting target or the implantation of particles within a surface are studied after exposure of the material to one or a number of discharges.

Langmuir probes are still one of the most commonly applied diagnostics in fusion devices as a relatively simple and inexpensive method to obtain time-averaged and fluctuating data from the edge plasma. The difficulty with direct plasma particle flux measurements is mainly in the understanding of the local perturbation of the plasma by the probe and of the relation of the local plasma parameters to the unperturbed plasma far from the probe. Different construction schemes are used to cope with the thermal load: (fast) moving probes [5], heat sink probes in good thermal contact with a big (cooled) heat sink; and flush mounted probes (in material surfaces which are exposed to the plasma at grazing incidence). The construction has to be compatible with the experimental demands; there is no optimal probe design suitable for every purpose. In Section III, the diagnosis of boundary plasmas will be illustrated, taking example by some types of advanced electrical probes developed on the tokamaks CASTOR and COMPASS at IPP in Prague (Czech Republic), and TEXTOR at Forschungszentrum Jülich (Germany) to measure plasma flows, T_e and T_i , and the plasma potential Φ_p.

3. Electrical probes

3.1. Plasma potential

Among the most important parameters is the electric space potential of the plasma which is usually called the plasma potential Φ_{pl}. Potentials can only be determined with respect to a certain reference potential, which is usually the wall. A Langmuir probe consists of a small electrode of various forms which is inserted into the plasma and externally biased with respect to the plasma potential. However, since the plasma potential is not directly accessible, the bias has to be applied with respect to the external reference electrode. A probe is called "cold" as long as it only passively registers the charge carrier fluxes towards it, but does not emit particles. The solid line in figure 1 shows a typical current-voltage characteristic, $I_p = I_p(V_p)$, of a cold probe in a conventional plasma, consisting of electrons and single-charged positive ions, where both particle species have Maxwellian velocity distributions [7]. At COMPASS it has been found [8] that in the vicinity of the inner and outer strike points of the divertor, the electron energy distribution can be approximated by a bi-Maxwellian with a dominating low-energy electron population (4-7 eV) and a minority of higher energy electrons (12-25 eV). The Langmuir probe characteristic is asymmetric, since the ion saturation current is much smaller than the electron saturation current, wherefore the floating potential V_{fl} of such a probe is more negative than Φ_{pl}. The reason for the strong discrepancy between the currents lies in the fact that the electrons have a much smaller mass than the ions and therefore a much higher mean velocity and average flux than those of the ions. The relation between the floating potential of a cold probe and the plasma potential in the case of a conventional Maxwellian plasma is well known from simple probe theory (T_e in eV):

$$\Phi_{pl} = V_{fl} + T_e \ln\left(\frac{I_{es}}{I_{is}}\right) \tag{1}$$

When the electron temperature T_e is known, it is in principle sufficient to measure the floating potential of a cold probe, and to calculate the plasma potential. However, on the one hand it is not so easy to measure T_e with sufficient accuracy, and on the other hand T_e can fluctuate during the measurement and there can be temperature gradients in the region of investigation (which is always the case in the edge region of a hot magnetized plasma). Moreover, also the ratio R between the ion and the electron saturation currents cannot always be determined precisely, especially in a strong magnetic field [7]. Furthermore, the entire characteristic of a cold probe shifts to the negative side and will therefore deliver erroneous results for the plasma potential whenever there is a stronger deviation of the electron velocity distribution function from a Maxwellian, e.g. in case of an electron drift or runaway electrons.

The two other characteristics in figure 1 show that the floating potential of a probe becomes identical to Φ_{pl} when the characteristic is symmetric. This can also be proven theoretically.

In principle there are two ways to achieve equal currents on both sides of the characteristic (i.e. R=1) and thereby a shift of the floating potential towards the plasma potential:

- Either compensate the plasma electron saturation current by an almost equally strong current on the negative (left-hand) side of the characteristic (dashed line in figure 1); see below: emissive probe.
- Or reduce the plasma electron saturation current on the positive (right-hand) side of the characteristic, until it becomes equal to the ion saturation current (dotted line in figure 1); see below: ball-pen probe.

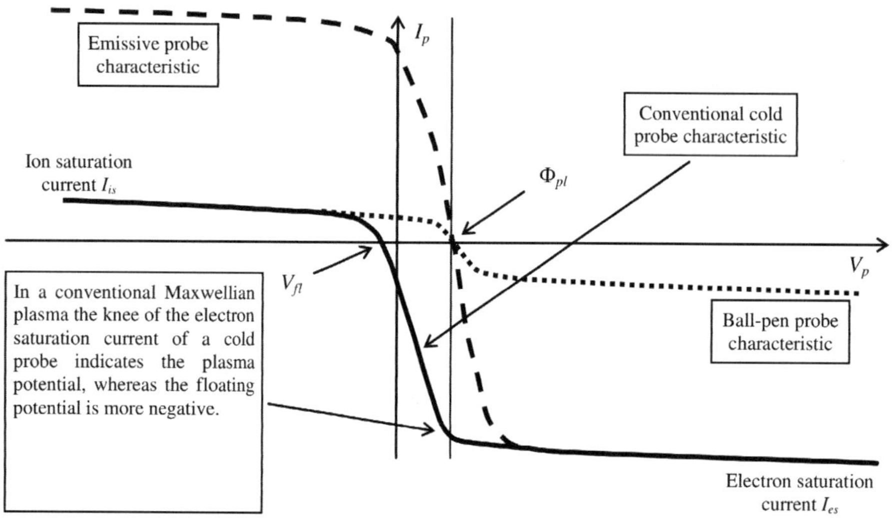

Figure 1. Typical current-voltage characteristics of a cold probe (solid line), of an emissive probe (dashed line), and of a ball-pen probe (dotted line) in a plasma with Maxwellian velocity distributions for the electrons and ions. Concerning the two latter probe types see below.

An electron emissive probe is usually realized by a small half-loop of tungsten or thoriated tungsten wire [7]. When such a probe is inserted into plasma and heated sufficiently, in the current-voltage characteristic the electron emission current Iem is observable on top of the ion saturation current I_{is} (cf. the dashed line in figure 1). This is due to the fact that a current of electrons emitted from the probe has the same sign as the current of positive ions flowing from the plasma to the probe. An emission current can flow as long as the bias of the probe is more negative than the plasma potential. By increasing the probe heating, the current on the left-hand side of the characteristic increases while the floating potential shifts to the right-hand side towards the plasma potential until a kind of saturation of this value is reached. Thus, when the emission current just compensates the electron saturation current (minus the ion saturation current, but this is usually negligible), the floating potential of such a probe equals the plasma potential: $V_{fl,em} = \Phi_{pl}$.

Up to now, emissive probes and heavy ion beam probes [9] were used for direct measurements of the plasma potential in tokamaks. However, the more widespread use of these techniques is hampered by various technical problems and peculiarities in the interpretation of the measured data. The heavy ion beam probe is a complex diagnostic, suitable for the measurement of the plasma potential in the interior plasma; moreover, its spatial resolution is limited. Directly heated emissive probes are rather fragile and of limited lifetime, and their exploitation on large fusion experiments is questionable. A laser heated emissive probe has been developed in view of fusion applications [10], allowing the use of materials like graphite for emissive wire probes, which are suitable for larger fusion devices.

An alternative approach to adjust R=1 which can be used only in magnetized plasmas, is the concept of the ball-pen probe (BPP) [11], approaching the principle of the Katsumata probe [12] (aiming at the measurement of the perpendicular ion energy distribution, and screening off electrons completely). The probe is designed so that the ratio R can be modified by changing the collecting areas for electrons and ions, taking advantage of the fact that the Larmor radii of electrons and ions are strongly different. The BPP has been developed on the small tokamak CASTOR (see schematic picture of figure 2). The probe consists of a conically shaped collector, which is shielded by an insulating tube made of boron nitride. The collector, which is movable inside the tube, is either completely shielded (h<0 in the left panel of figure 2) or partially exposed (h>0 in the right panel of figure 2) to the plasma. The probe collector can be biased by a swept voltage to estimate T_e and the ratio R from the I-V characteristics, or can be electrically isolated (floating) to measure the potential

V_{probe} and its fluctuations without the need of a power supply. In the ideal case, when the collector is hidden inside the tube, in principle only ions with sufficiently large Larmor radius can reach the collector surface, and the collecting area for electrons is zero. Consequently, the ratio $R = 0$. When the collector is moved outwards the electron current as well as R increase. At a certain collector position, the electron and ion currents are expected to be balanced (i.e., $R = 1$).

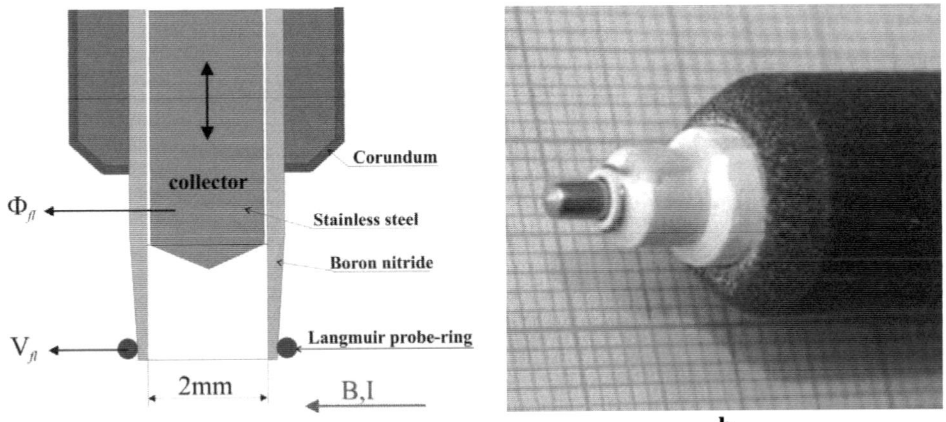

Figure 2. (a): Schematic of the ball-pen/standard Langmuir probe combination. The conical collector can be moved inside the boron nitride (BN) screening tube that acts as a shield for electrons, h is the position of the collector tip inside the tube relative to the top cross section plane of the shielding BN tube. The Langmuir probe is made of 0.2 mm diameter tungsten wire. (b): photograph (fully exposed collector).

The results of systematic BPP measurements of the floating potential V_{fl} and the ratio R on the collector position are plotted in figure 3-right. It is evident from the figure that ln(R) is always positive for any collector position. It indicates that electrons are present in the shadow of the shielding tube, and that the electron current is always higher than the ion current. This is in contrast with the simple model based on the electron and ion gyromotion along the magnetic field lines, which is described above. Nevertheless, ln(R) attains a minimum (i.e. ln(R)=0.1), when the tip of the collector is slightly inside the shielding tube ($h = -0.5$mm). In this situation the probe potential is close to the plasma potential. According to (1) the difference between plasma and probe potential is in order of volts in this case ($T_e \approx 10$eV). Although the electrons can in principle enter the BPP tunnel, it is assumed that a part of them with favorable of the velocity components parallel and perpendicular to the magnetic field, can bypass the tunnel. As such the tunnel works as a filter in velocity phase space, deforming the Maxwellian distribution of particle velocities, which is one of the assumptions of Langmuir probe theory. However, 2D cartesian Particle-In-Cell simulations [13] of a simplified model of the BPP have verified the validity of (1). When R is close to unity, the floating potential of the collector corresponds to the plasma potential obtained by the analysis of the first derivative of the IV characteristics.

It is seen from figure 3 that the value of the probe potential significantly decreases when the collector is more and more exposed to the plasma. When the collector is fully outside the shielding tube ($h \approx 1.5$mm), the probe operates as a conventional single Langmuir probe and measures the floating potential V_{fl}^{O}. It is interesting to note, that the probe potential is approximately constant and equal to the plasma potential Φ, when the collector is hidden inside the tube. The reason for this behaviour is not clear. However, it has practical importance for the direct plasma potential measurements by this probe, because the collector can be hidden at any position inside the shielding tube deeper than in the previous case ($h = -0.5$mm) and protected against high energy flux. Therefore, such a robust Katsumata-type probe [14] with flat collectors has been used on the larger devices RFX (reversed field pinch) in Padova and on ASDEX Upgrade in Garching. On ASDEX

Upgrade a quadrupole BPP (probe head with 4 BPPs with different retraction depth $h = -0.5$mm, -1mm,-2mm, and-3mm) mounted on a fast reciprocating manipulator has been used for direct measurements of the edge plasma potential in ELMy H-mode discharges [15]. It was found that all the BPPs measure the radial profile of the plasma potential independently of the position of their collector, which is in good agreement with measurements on CASTOR.

The credibility of the BPP has been proved not only with emissive probe and self-emitting Langmuir probes during deep reciprocation on COMPASS and ASDEX Upgrade, but also the Doppler reflectometry measuring radial electric field on ASDEX Upgrade [16]. Its capability of resolving both plasma potential and electron temperature on the turbulence (microsecond) time scale was demonstrated [17] with results consistent with a model based on SOL interchange-driven turbulence transporting both energy and particles by blobs. In addition, in the limiter shadow, BPP seems to yield more credible T_e (sharply decreasing down to 1eV), where swept Langmuir probes show improbable and constant T_e~10eV.

3.2. Electron and ion temperature

Among the various diagnostic tools to measure the electron temperature T_e in a plasma, probes are the least expensive, simplest and most versatile. Probes can be used in most types of plasma, even in the edge plasma of medium-size tokamaks, and they allow localised measurements with good spatial resolution. The usual and best known method to determine T_e with a cold probe is to register the current-voltage characteristic (see figure 1) and to evaluate the exponential increase of the electron current in the retarding field region. A disadvantage of this method is its low temporal resolution which naturally is limited by the frequency with which the characteristic can be scanned. In principle also the difference between the floating potential V_{fl} and the plasma potential Φ_{pl} (1) contains the value of the electron temperature.

The Langmuir probe-ring attached around the BN tube of the ball-pen probe (figure 2) is fully exposed to the normal flux of electrons and ions from the plasma. Therefore, its floating potential corresponds to the conventional value. Thus the difference between the floating potentials of the correctly adjusted ball-pen probe (for $h \cong -1$ mm where $R(h) = 1$) and that of the Langmuir probe-ring can be used for calculating T_e [18] according to (1).

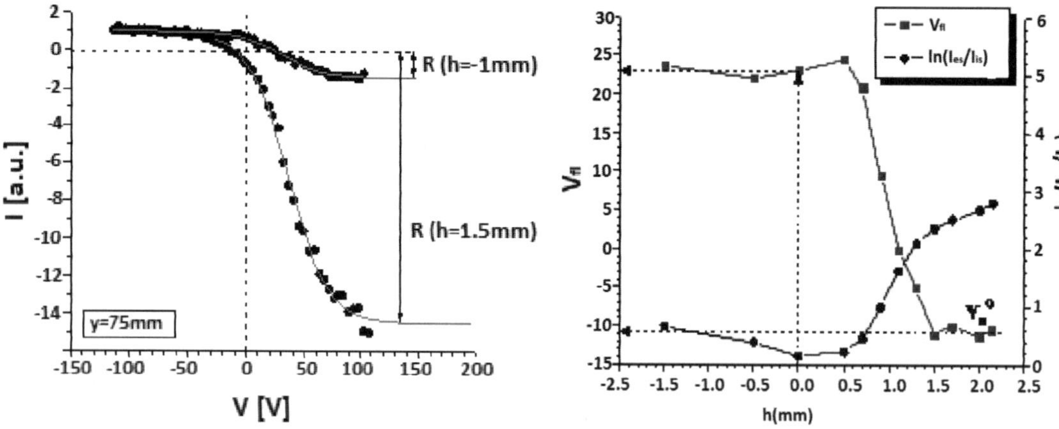

Figure 3. (a) -Example of I-V characteristics of the BPP in CASTOR for two different collector positions. The h value is negative when the collector is hidden inside the shielding tube. (b) - The variation of the floating potential V_{fl} and $\ln(R)$ with respect to the collector position. The probe head is at plasma radius 75 mm.

A new kind of robust Langmuir probe, the (segmented) tunnel probe (TP) has been developed for measurements of the electron and ion temperature [19, 20, 21].The probe is shown in figure 4 and consists of a hollow conducting tunnel (TUN) of 5 mm diameter that is closed at one end by an electrically isolated 5 mm diameter conducting back plate (BP). Both conductors are biased negatively to collect ions and repel electrons. The tunnel axis is parallel to the magnetic field B. Plasma flows into the orifice, and the ion flux is distributed between the tunnel and the BP [20]. The ratio $R_c = I_{TUN} / I_{BP}$ between the current to the tunnel (two segments shorted together) and that to the back plate, respectively, is determined by the thickness λ_{MS} of the magnetic sheath (MS) at the concave surface of the tunnel, and is therefore a strong function of T_e. The physics governing the TP is fundamentally different from that of a classical Langmuir probe (LP). The applied voltage on the LP is swept in order to measure a restricted part of the electron distribution function. The TP, on the other hand, is biased to a fixed potential that is sufficiently negative to repel all electrons. The temperature of the electrons is measured even though none are collected. The ion current for highly negative voltages is almost perfectly saturated, demonstrating the immunity of concave probes to sheath expansion effects, a problem that has always plagued convex LP applications [2,6]. Because it is concave, the electric field is contained inside the TP independent of the applied voltage. Furthermore, being oriented in the parallel direction, the TP does not collect the E×B component of the flow. The TP can operate in DC mode and therefore provides fast measurements of T_e (fluctuations), which is very important to understand transient heat transfer. Moreover, due to the clearly defined tunnel orifice, this probe is not subject to the uncertainties of collecting area from which classical convex probes suffer [2,6]. The tunnel radius should be roughly twice the thickness of the magnetic sheath. Precise self-consistent, two-dimensional kinetic "particle-in-cell" code XOOPIC [20] modelling is possible because of the simple TP geometry (classical Langmuir probes require extensive 3D simulations), and is used to determine the theoretical relation between the current ratio $R_c = I_{TUN} / I_{BP}$ and T_e. Qualitative agreement with classical LP measurements is found, but the electron temperature given by the TP is several times lower [22]. It has been found that this discrepancy can be caused by secondary electron emission. It suffices to correct the BP current assuming some reasonable value for the secondary electron emission coefficient. Suprathermal electrons can also contribute to the discrepancy since a swept LP is sensitive to it [23].

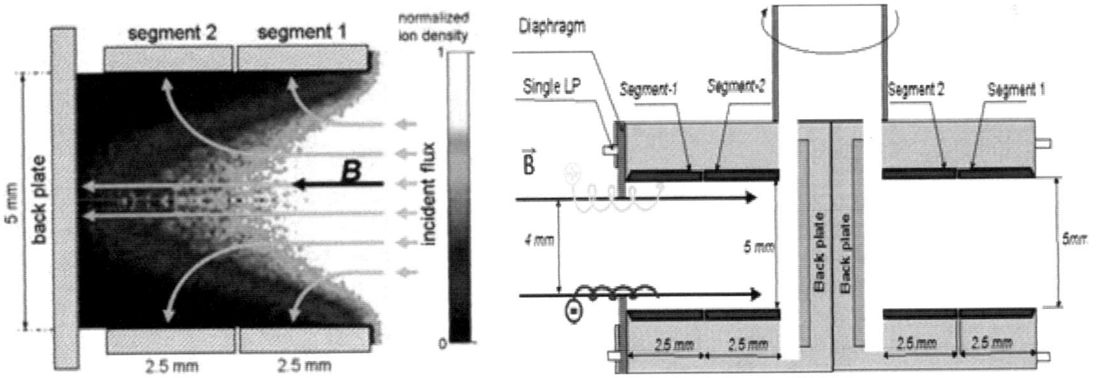

a **b**

Figure 4. (a): schematic of the segmented tunnel probe (TP). The current collected by each of the three electrically insulated conductors is monitored separately. The ion guiding centre trajectories are shown by arrows. (b): modified TP; left hand side: ion-sensitive segmented Katsumata TP; right hand side standard segmented TP.

The ion temperature in the tokamak scrape-off layer is notoriously difficult to measure and thus rarely available. One of the most commonly applied probes for T_i measurements in the SOL is the retarding field analyzer (RFA) [6]. Unfortunately, since the RFA is working in a swept mode it does not allow the measurement of T_i with a sampling rate faster than a few milliseconds and thus cannot contribute to the measurement of fluctuations and transient phenomena like ELMs. Therefore, the development of a new probe diagnostics for T_i measurements in the SOL is an important issue.

The ions that flow into the orifice of the segmented TP (figure 4) are diverted onto the tunnel surface by the intense radial electric field in the magnetic sheath. XOOPIC simulations have shown that the axial distribution of ion flux onto the tunnel decays with a characteristic scale length that is determined by the relative strength of the radial acceleration of ions with respect to the incident parallel ion velocity. The latter is a function of the ion sound speed Therefore, by splitting the tunnel into two electrically insulated segments, the (parallel) ion temperature T_i can be found [24] from the ratio of ion flux to the first and the second segment, $R_c = I_{seg1} / I_{seg2}$.The advantage is that the TP is operated in DC mode and thus provides fast measurements of (parallel) T_i as well as of the parallel ion current density $J_{\|i}$.

The modified TP shown in the right panel of figure 4 allows to measure the perpendicular ion temperature [19]. In front of the entrance orifice of the tunnel an additional diaphragm can be mounted (at left-hand side only in figure 4), which transforms it into an ion-sensitive (Katsumata) probe, the modified Katsumata tunnel probe. Due to the diaphragm, which protrudes from the tunnel by 0.5mm around the entire circumference, the electrons should in principle not be able to reach the tunnel segments at any potential of all electrodes since their gyroradii are in the range of 20μm. The back plate is left floating, and the voltage on the diaphragm and both tunnel segments is swept simultaneously [24]. The segmented TP has reached the level of technical performance at which it starts to provide a wide range of useful scientific results and can be experimentally tested in larger fusion devices [19].

4. Magnetic probes

In the plasma interior magnetic fields are measured mostly using spectroscopic methods. However, particularly on small and medium sized fusion devices, magnetic probes are applied to study magnetic turbulence at the plasma edge. For this purpose standard pick-up coils as well as Hall sensors are used. Additionally, miniature Rogowski coils were successfully applied to measure edge current density profile and its fluctuations on CASTOR [25] and RFX [26]. From magnetic field measurements performed outside the plasma important properties like plasma current, energy content and MHD fluctuations together with their mode structure can be inferred. Such measurements utilize different types of coils. As the discharges became longer, the evaluation of B from its measured time derivative has become increasingly difficult, because the integration needs a precise determination of possible offsets in the preamplifiers [27].

Advancements in semiconductor technology hand in hand with a broad spectrum of industrial applications have driven development of new types of Hall sensors for magnetic measurements in recent years [28]. These sensors with their simple principle of operation and linear output voltage dependence on magnetic field are well established for many applications in experimental physics as well as in industry. A particular advancement is the availability of 'integrated' Hall transducers, where the sensing element together with the complex electronic circuitry is integrated on a single small chip with characteristic dimension of a few millimeters. The on-chip integrated circuits provide stabilization of the supply voltage, output voltage amplification, signal conditioning in order to suppress the high frequency noise, and elimination of temperature dependence of the sensor's output. In particular, the output amplifier placed directly next to the sensing element significantly improves the frequency response and the signal to noise ratio. Because of the widespread industrial use of such sensors, their cost is rather low. Recently, Hall probes have been used to measure the absolute value of B directly together with its fluctuations in the boundary plasma of tokamaks [28, 29]. The strongest objection against the use of Hall sensors in the reactor type tokamak is their assumed vulnerability to radiation damage (in particular to the high neutron fluxes) [27]. The first results of irradiation tests of recently developed 'radiation-hard' Hall sensors [30, 31] were encouraging. The sensitivity of metal Hall sensors based on copper showed good agreement [32] with theoretically expected values and remained constant in the ITER expected temperature range between 100 and 250°C. Post irradiation characterization of samples irradiated up to a total neutron fluence of $2 \cdot 10^{18}$ cm^{-2} showed no effect on the sensitivity [32].

References

[1] P.C. Stangeby and G.M. Mc Cracken, 1990 *Nucl.Fus.* **30**, 1225
[2] I.H. Hutchinson 2002 *Principles of Plasma Diagnostics* Cambridge University Press
[3] G. Van Oost 2008 *Transactions of Fusion Technology* **53 2T** 356-366
[4] G. Van Oost et al. 2007 *Nucl. Fusion* **47** 378
[5] J. Boedo et al. 1998 *Rev. Scient. Instr.* **69** 2663
[6] G.F. Matthews 1994 *Plasma Phys. and Contr. Fus,* **36** 1595
[7] R. Schrittwieser et al. 2002 *Plasma Phys .and Contr. Fusion* **44** 567
[8] M. Dimitrova et al. 2014 *Contrib. Plasma Physics* **54** 255
[9] All Papers in 1994 *IEEE Transact. On Plasma Sci.* **22, No. 4**
[10] C. Ionita et al. 2010 *Contrib. Plasma Phys.* **51** 264
[11] J. Adamek et al. 2004 J., *Czechoslovak Journal of Physics* C **54** 95
[12] I. Katsumata and M. Okazaki 1967 *Japan J. Appl. Phys.* **6** 123
[13] M. Komm et al. 2010 *Contrib. Plasma Phys.* **50** 814
[14] R. Schrittwieser et al. 2006 *Czech. J. of Phys.* **56** B145
[15] J. Adamek et al. 2009 *J. Nucl. Mater.* **390-391** 1114
[16] J. Müller et al. 2011 *Nucl.Fus.***51** 073023
[17] J. Horacek et al. 2010 *Nucl.Fus.* 105001
[18] J. Adamek et al. 2005 *Proc. 32th EPS Conf Plasma Phys. Contr. Fus.* (Tarragona, Spain) P5.081
[19] J.P. Gunn et al. 2002 *Czech. J. of Phys* **52** 1107
[20] J.P. Gunn et al. 2004 *Rev. Scient.Instr.* **75** 4328
[21] M. Kocan et al. 2007 *J. Nucl. Mater.* **363-365** 1436
[22] J.P. Gunn et al. 2004 *Czech. J. of Phys* **55** 255
[23] T. Van Rompuy et al. 2007 *Plasma Phys. and Contr. Fusion* **49** 619
[24] P. Balan et al. 2004 *Contrib. Plasma Physics* **44** 683
[25] J. Stöckel *et al. 1999 Plasma Phys. and Control. Fusion* **41** A577
[26] E. Torbert *et al. 2003 Rev. Scient.Instr.***74** 5097
[27] G. Vayakis, C. Walker and the ITER International Team and Participant teams 2003 *Rev. Scient.Instr.* **74**, 2409
[28] I. Duran et al. 2002 *Rev.Scient.Instr.***73** 3482
[29] I. Duran et al. 2008 *Rev. Scient. Instr.* **79** 10F123
[30] I. Bolshakova et al. 2007 *Sensor Letters* **5** 283
[31] I. Duran et al. 2010 *Rev. Scient.Instr.* **81** 10E122
[32] J. Sentkerestiova et al. 2013 *Fusion Engineering and Design* **88** 1310

CXRS measurements of ion temperature in NBI discharges on Globus-M spherical tokamak

G F Avdeeva[1,3], **I V Miroshnikov**[2], **N N Bakharev**[1], **G S Kurskiev**[1], **M I Patrov**[1], **V Yu Sergeev**[2] **and P B Schegolev**[1]
[1] Ioffe Institute, Politekhnicheskaya 26, St Petersburg, 194021, Russia
[2] Peter the Great St Petersburg Polytechnic University SPbPU, Polytechnicheskaya 29, St Petersburg, 195251, Russia

E-mail: galina2791avdeeva@gmail.com

Abstract. The thesis describes the Charge-eXchange Recombination Spectroscopy (CXRS) diagnostics setup and its application to the ion temperature determination in the NBI-heated discharges in the Globus-M spherical tokamak. Measurements of the spectral line shape of the radiation emitted by the hydrogen-like carbon ion CVI (5290.5 Å) are presented. Comparison of CXRS and Neutral Particle Analyser (NPA) data demonstrates good agreement.

1. Introduction

The temperature of the ion component is one of the key physical parameters of a high-temperature plasma. Data on the spatial distribution of the ion temperature are necessary for understanding and modeling of the particles and energy transport and for assessment the efficiency of plasma heating systems. Thus, the ion temperature diagnostic is one of the most important components of a diagnostic complex of modern device for magnetic confinement of plasmas [1-3].

Two methods of measuring the ion temperature are available on the Globus-M tokamak: Charge eXchange Recombination Spectroscopy (CXRS) and Neutral Particle Analyzer (NPA) diagnostic [4]. The CXRS is based on the charge-exchange reaction of impurity nuclei on the neutral atoms of a diagnostic or heating beam, produced by the neutral beam injection (NBI). Hydrogen-like ions are generated when a beam of fast neutral hydrogen particles is injected into the plasma, that results in the recharge of residual nuclei on the beam particles. If the generated ion is in an excited state, a transfer to the ground state occurs through a cascade of transitions with the emission of the corresponding spectral lines. The most convenient is the use of transitions within the range of the visible light. The energy transfer time between hydrogen particles in the main plasma and residual ions is substantially smaller than the energy confinement time. Therefore, the temperature of the impurity ions, which is determined by the CXRS diagnostic, coincides with that of ions of the main plasma. The CXRS method allows to measure the local values of plasma temperature and to obtain the temperature profile using the multichord measurements.

[3] To whom any correspondence should be addressed.

2. Experimental setup

Measurements were performed on the Globus-M spherical tokamak with a major radius $R = 36$ cm and a minor radius $a = 24$ cm [5]. In the analyzed discharges the average value of the plasma density was $n \sim 3\text{-}5 \cdot 10^{13}$ cm^3, the toroidal magnetic field $B_{tor} \sim 0.4$ T, plasma current $I_p \sim 115\text{-}200$ kA. Typical ion temperature in the ohmic discharges is 200 eV. The beam of neutral atoms was injected in the tangential direction and had a transverse dimension (level e^{-1}) about 3 cm × 10 cm. The beam energy was 23-28 keV, current, 22-28 A. There were two viewing chords: first chord intersects the NBI axis at $R = 40$ cm; the second, at $R = 45$ cm (figure 1).

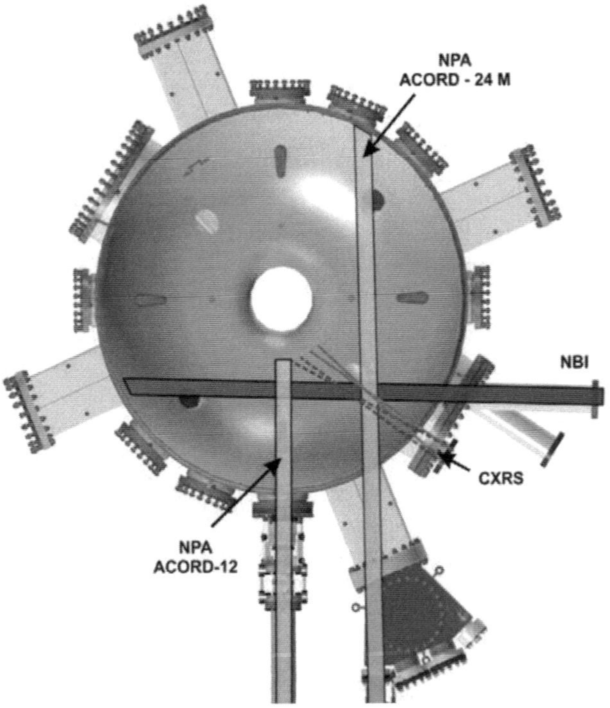

Figure 1. Experimental setup. Full line shows a field of the CXRS viewing chord which intersects the NBI axis at $R = 45$ cm, dotted line, at $R = 40$ cm; NPA ACORD-24M observation chord intersects the NBI axis at $R = 45$ cm, ACORD-12, at $R = 32$ cm.

Two 15 m long light optical fibers transferred the radiation from the installation room to the location of measuring equipment. After passing through a simple optical system, the light from both fibers is focused on the entrance slit of the spectrometer. The images were separated by height and did not overlap. A diffraction high-resolution spectrometer [6] operates on the principle, based on the double diffraction at grazing incidence of light on the grating. The device has the input and output lenses with a focal length of $F = 300$ mm, the grating 1800 grooves / mm, inverse linear dispersion of 1.5 Å / mm and relative aperture F/7, the working range of 4300 - 9000 Å. The spectrum was registered with the EMCCD camera «Andor iXon Ultra», having a high quantum efficiency (up to 90% at a wavelength of 5400 Å) and the low values of noise signal (at the level of units of electrons). The optical scheme is shown in figure 2.

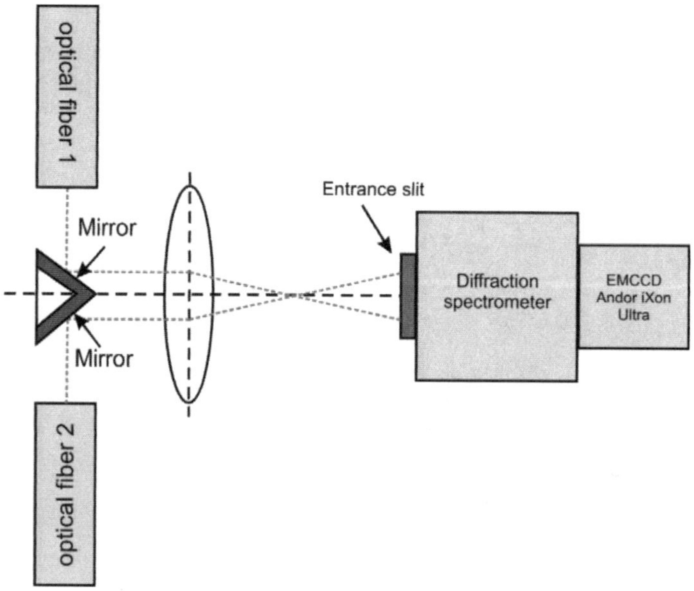

Figure 2. Optical scheme.

3. Processing method and the obtained results

To determine the values of the ion temperature the spectral line shape of the emission of a hydrogen-like carbon ion CVI (5290.5 Å) was registered in the discharges in the Globus-M tokamak with the NBI heating. To improve the signal/noise ratio, the raw experimental data for each moment of time were averaged in a series of discharges.

Each registered spectral line shape is an integral along the line of sight of each observation chord. Since plasma contains all kinds of ions for each kind of impurity, the radiation at the same wavelength enters the line of sight of the spectrometer from the edge of plasma. Accordingly, to determine the ion temperature from the measured spectra an approximation was used which assumes the spectrum to be a sum of the background (Bremsstrahlung radiation) and two Gaussians: a "passive" component of the signal coming from the plasma edge and an "active" charge exchange component caused by the neutral beam:

$$F = F(a_k, \lambda) = Backgr + F_{active} + F_{passive} =$$

$$Backgr + a_{active} \cdot \exp\left[\frac{-m \cdot c^2}{2 \cdot Ti_{active}}\left(\frac{\lambda}{\lambda_0} - 1 + \frac{V_{active}}{c}\right)^2\right] + a_{passive} \cdot \exp\left[\frac{-m \cdot c^2}{2 \cdot Ti_{passive}}\left(\frac{\lambda}{\lambda_0} - 1 + \frac{V_{passive}}{c}\right)^2\right],$$

where the values for 7 coefficients, namely $Backgr, a_{active}, Ti_{active}, V_{active}, a_{passive}, Ti_{passive}, V_{passive}$, were recovered using the Levenberg-Marquardt method for every spectrum.

The effect of the fine structure and Zeeman splitting leads to overestimated values of temperature from CXRS [7] by 15% for the central plasma and 30% for the edge under Globus-M conditions. In the present results this effect is taken into account.

Figure 3 shows the results of processing the spectra and comparing with the NPA data in one of the researched regimes. The CXRS exposure time was 5 ms, that of the NPA, 9 µs. The obtained values of the ion temperature are 250-370 eV. Given the fact that these diagnostics provide local temperatures at the points where R = 32 cm (NPA), R = 40 cm and R = 45 cm (CXRS), their results do not contradict each other, and the temporal variation of temperature are in a sufficiently good agreement.

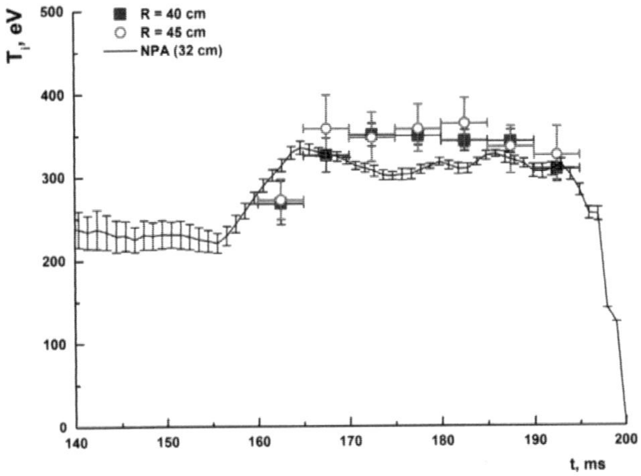

Figure 3 Comparison of ion temperature values obtained with the CXRS and NPA (ACORD-12) diagnostics in a regime with the plasma electric current I = 200 kA.

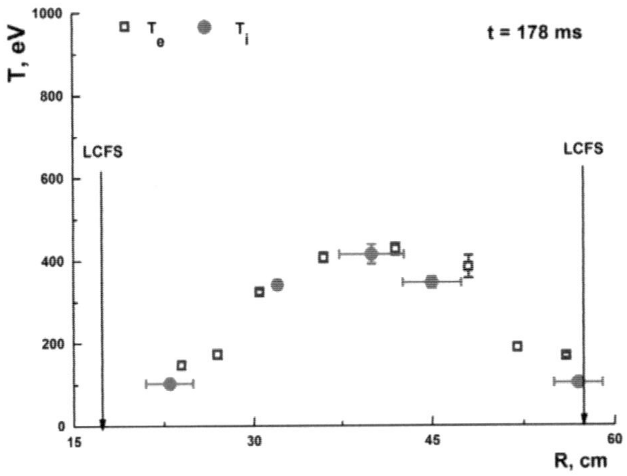

Figure 4 Ion and electron temperature profiles in a regime with the plasma electric current I = 200 кA. Empty markers show the electron temperature, full markers, the ion temperature. Vertical arrows indicate the separatrix position.

Figure 4 shows the profile of the ion temperature compared to the profile of the electron temperature (obtained by the Thomson scattering diagnostics) at the same time instant. One can see that by the end of the discharge (at 178 ms) the values of ion and electron temperatures are almost equal, and the profiles have the same shape.

4. Conclusion

The measurements of the spectral line shape of the radiation emitted by the hydrogen-like carbon ion CVI (5290.5 Å) were performed in the discharges in the Globus-M tokamak with the NBI heating. Comparison of the obtained results for ion temperature from the CXRS and NPA diagnostics demonstrates a good agreement.

5. Acknowledgement

This work was conducted using the equipment of the North-Western Common Use Center "Matter Research and Diagnostics in High Technologies" with financial support of the Ministry of Education and Science of the Russian Federation (Agreement № 14.621.21.0007, identifier RFMEFI62114X0007).

References

[1] Yoshinuma M Yokoyama M, Osakabe M and Nagaoka K 2010 *Fusion Sci. and Technology* **53** 375
[2] Viezzer E, Pütterich T, Dux R and McDermott R M 2012 *Rev.Sci.Instrum* **83** 103501
[3] De Bock M et al 2004 *Rev.Sci.Instrum* **75** 4155
[4] Bakharev N N et al 2013 *Tech. Phys. Lett.* **39** 1085
[5] Gusev V K et al 2013 *Nuclear Fusion* **53**. 093013
[6] KoshkinYu S et al 1999 *Rev. Sci. Instrum.* **70**. 3844
[7] Blom A and Jupen C 2002 *Plasma Phys. Control. Fusion* **44** 1229

Multifunctional material probe for impurity fluxes characterization

L Begrambekov[1] and A Grunin
National Research Nuclear University MEPhI (MEPhI), Kashirskoe shosse 31, Moscow, 115409, Russian Federation

E-mail: lbb@plasma.mephi.ru

Abstract. A new device is proposed for diagnostics of the SOL plasmas in tokamaks. The device allows to obtain a more detailed information on the impurity concentration, ionization state distribution, and particle fluxes, by combining the several techniques in a single device. The operation principle and the main parts of the proposed multifunctional probe are described.

1. Introduction

The usage of material probes in plasma and thermonuclear devices allows obtaining the information on the plasma energy, concentration, the chemical composition and the amount of impurity in the scrape-off-layer (SOL) plasmas, among other benefits. The practice of employing the probes in the tokamaks has shown that the value of the information obtained via probing increases dramatically if the probe is able to register several parameters, rather than single one, of the SOL plasma to give the complex information on the SOL plasma properties, including the ionization state distribution, chemical composition and the amount of impurity.

2. Problem setting

Currently, such probes do not exist, and its development is the goal of this work. The probe in question will enable one to implement the following measurements and researches:

- In situ measurement of the total flux of neutral and charged impurity particles;
- Research of the interaction of the SOL plasma with the samples of the first wall material and construction materials in tokamaks;
- In situ measurement of the total flux of ions;
- In situ measurement of the mass components of the ion flux;
- In situ evaluation of the energy spectra of the hydrogen isotope ions in the SOL plasma.

The working elements of the probe are located in the protecting jacket which screens the sections from the plasma particles and the heat flux. The jacket is perforated to allow the plasma particles to reach the sections where the following measurements can be made: mass-spectrometry, hydrogen ion energy analysis, in situ neutral and charged particles flux analysis, and analysis of deposited and trapped plasma particles.

The components of plasma passing through the holes in the jacket are accumulated in the plasma components deposition section either on the substrate surfaces (solid material impurities) or

[1] To whom any correspondence should be addressed.

accumulated in the deposition layer (gas ions and atoms). After removing the substrate from the probe the deposited layer is analyzed using the SIMS, RBS, AES and other methods. Trapped plasma components are analyzed with the TDS technique. The minimum detectable amount of the trapped atoms is $5\cdot10^{14}$ at/cm^2. The sensitivity and the resolution of the device during the analysis of the trapped and deposited impurity depends on the specific method of the analysis and the respective device used.

The in situ neutral and charged particle flux analyzer collects the particles on a thin layer of a conductor with known resistance. After the particles are deposited, the value of the resistance is changed, that is then registered. The resolution of the registration of impurity flux varies in the range from one deposited layer (about 10^{15} atoms/cm^2), with the total number of deposited layers being less than a hundred on the surface of the plate, up to five layers (about $5\cdot10^{15}$ atoms/cm^2), with the total number of deposited layers lying between one hundred and five hundreds.

3. Discussion

In the mass-spectrometer (figure 1) the ions which enter the probe through the hole in the jacket are accelerated in the region between the jacket and the biased diaphragm. These ions are then mass-analyzed in the crossed E×B fields, where B is the tokamak's magnetic field and E is the electric field between the plates of the probe's electric field analyzer. Afterwards, the particles are registered on their respective collector. The entire section is biased and has the potential of the accelerating diaphragm. In figure 1, one can see the schematic of some elements of the mass-spectrometer which is now being constructed for the WEST tokamak with magnetic field of 4T.

The deviation from the initial trajectory of the ions depends on the energy and the mass-to-charge ratio (m/Z). The mass-spectrometer is able to register ions in a wide range of the values of the m/Z parameter (between 1 and 200) and is limited by the value of potential biasing of the electric field plates. Preacceleration of the ions allows to minimize the separation of the ions with the similar values of the m/Z ratio and various values of the energy. Thus, the acceleration of the ions with the energy gain of the order of 500 eV, according to calculations, will allow to detect ions in the energy range from 1 eV to 100 eV. The registration resolution is 10 for the light ions (m/Z~10) and 3-5 for the heavier ones (m/Z~100).

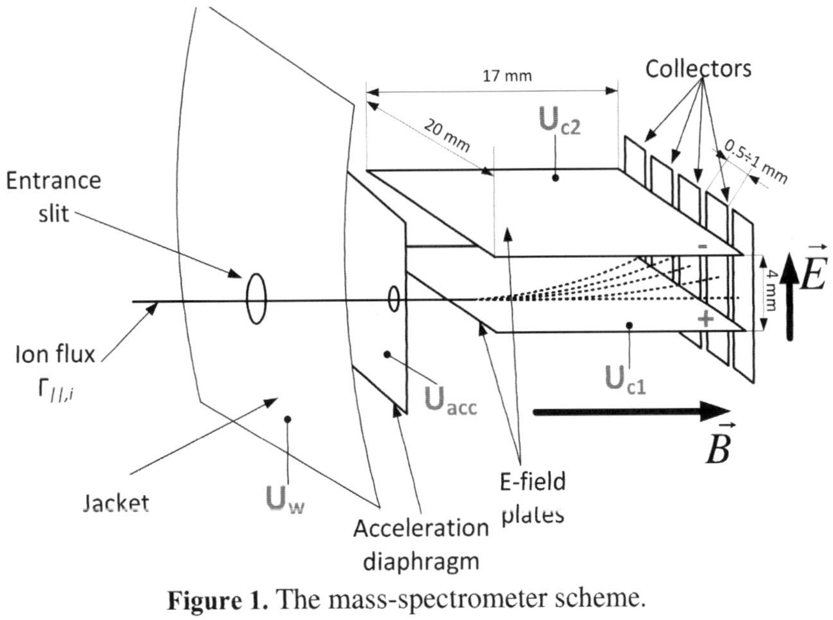

Figure 1. The mass-spectrometer scheme.

Figure 2 shows a simulation of the ion motion for different values of the m/Z ratio (9, 56, and 189). It is seen that the ions with different values of m/Z are collected at different collectors, creating the current in the circuit. The minimum detectable concentration of impurity (bounded by the condtions of measuring the ion current onto collectors) is 0.01 at.% for ions with the ratio m/Z~100 and 0.001 at.% for ions with m/Z~10. In this case, the currents in the collector circuit will be at least 0.1 μA.

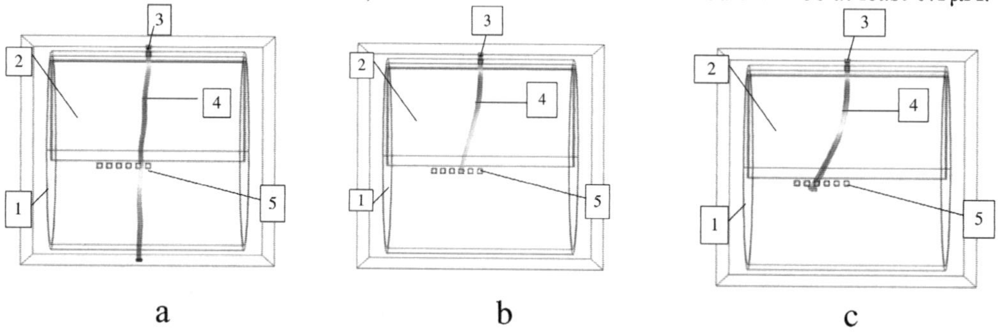

Figure 2. The trajectories of particles with different values of the m/Z ratio: 9 (a), 56 (b) and 189 (c). 1 - jacket, 2- E-field plates, 3- entrance slit, 4 - ion trajectories, 5 - collectors.

The hydrogen ion energy analyzer is structurally similar to that of the section shown in figure 1, but has no accelerating electrode. The section is able to detect ions of hydrogen and deuterium in the energy range of 1÷500 eV. Energy resolution ($E/\Delta E$) equals 10. The minimum detectable flux of ions is $1\cdot10^{13}$ ion/s.

The distance between the electric field plates is taken such that the ions with the parameter m/Z> 2 does not reach the collectors. The trajectories of H+ ions of different energies are shown in figure 3.

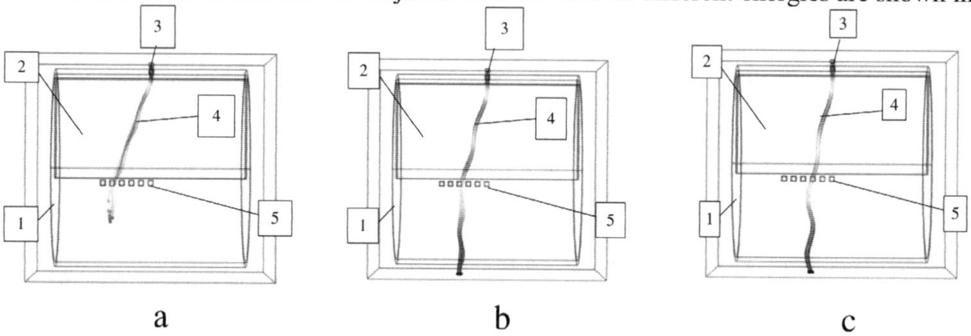

Figure 3. The trajectories of hydrogen ions with energies 50eV (a), 150eV (b), 300eV (c).
1 - jacket, 2 - E-field plates, 3- entrance slit, 4 - ion trajectories, 5-collectors.

Figure 4 shows the frame on which the five sections of the measuring probe are mounted (they are shown in figure without jacket). Its main elements are the two parallel titanium plates fastened with ceramic bridges. Three sections of collecting neutral and charged impurity particles and of accumulation of deposited/implanted ions from the plasma components (with the six substrates placed here), mass spectrometry section, and the section of in-situ measuring the neutral and charged particle fluxes, are installed in-between titanium plates. The frame will be installed in a protective cylindrical jacket, with inner diameter of 35 mm. The overall dimensions of the probe are 250mm × Ø40mm.

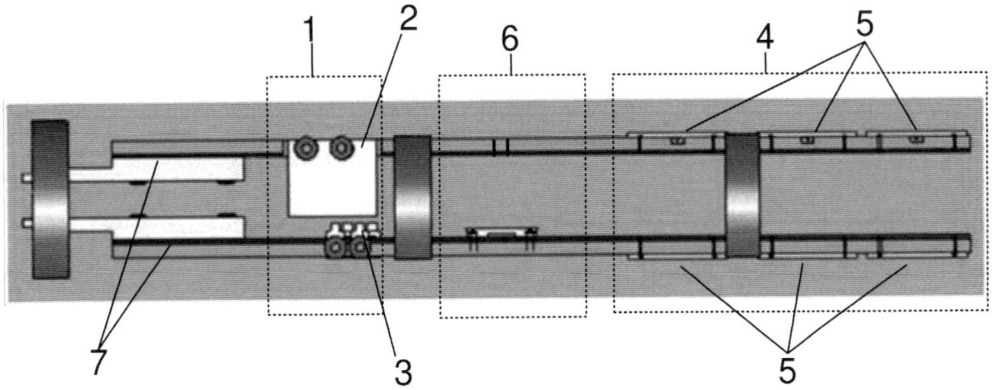

Figure 4. The model assembly of the probe. 1 - Mass-spectrometer, 2 - E-field plates, 3 - collectors, 4 – ion and atom accumulator section, 5 - substrates, 6 - in situ neutral and charged particle flux analyzer.

References

[1] VanOost G 2004 *Fusion Science and Technology* **45** 301-308
[2] Rubel M, Wienhold P and Hildebrandt D 2003 *Vacuum* **70** 423-428
[3] Rubel M, Coad J, Wienhold P, Philipps V, Stamp M and Tanabe T 2004 *Phys. Scr.* **T111** 112-117

Journal of Physics: Conference Series **666** (2016) 012004 doi:10.1088/1742-6596/666/1/012004

A method of interpreting the Balmer-alpha high-resolution spectroscopy for tokamak edge plasmas with account of divertor stray light

V S Neverov[1,3], A B Kukushkin[1,2] and A G Alekseev[1]
[1] National Research Center "Kurchatov Institute", Kurchatov square 1, Moscow, 123182, Russia
[2] National Research Nuclear University MEPhI, Kashirskoe shosse 31, Moscow, 115409, Russian Federation

E-mail: vs-never@hotmail.com

Abstract. A method is suggested for interpreting the data from the Balmer-alpha high-resolution spectroscopy diagnostics of the edge plasma in the tokamak main chamber, which additionally uses the data from direct observation of the divertor. Such an extension of the diagnostics is motivated by the fact that in a tokamak-reactor with the metal first wall, like ITER tokamak, a significant role of the divertor stray light (DSL), which is emitted by the plasma in the divertor in the same spectral line and reflected from the first wall of the vacuum chamber to a spectrometer in the main chamber, is expected. The results of the first applications of the developed model to interpret the data from the JET-ILW tokamak experiments, which simulate the conditions of occurrence of the DSL in ITER, are discussed.

1. Introduction

The Balmer-alpha high-resolution spectroscopy in the ITER tokamak, aimed at measuring the characteristics of hydrogen isotopes in the scrape-off layer (SOL) plasma, may face the problem of the divertor stray light (DSL), which is emitted by the plasma in the divertor in the same spectral line and reflected from the first wall of the vacuum chamber to a spectrometer in the main chamber. Preliminary estimates [1] of the DSL spectrum and the numerical modeling [2] (using the ray tracing technique) of the DSL, not resolved within spectral lines shape, for the ITER Main Chamber H-alpha diagnostics have shown that the DSL in the Balmer-alpha line may exceed the Balmer-alpha light, emitted in the SOL on the observation chord in the main chamber, up to two orders of magnitude for highly reflecting walls (wall reflection coefficient $R_w >= 0.5$) and high-power operation. The results [1, 2] have been obtained with the use of the spatial distribution of luminosity in the divertor and the SOL for the quasi-stationary stage of the inductive mode of ITER operation with the fusion gain parameter, Q=10 (these data were simulated by the SOLPS4.3 (B2-EIRENE) code [3-5] on an expanded numerical mesh with allowance for the poloidally resolved recycling from the first wall [6]). The first results [1] showed that a test of the elaborated approach in the currently running machines with all-metal first wall is required to benchmark the analysis method.

Here we propose a method of interpreting the data from the Balmer-alpha high-resolution spectroscopy of the edge plasma in the main vacuum chamber of a tokamak, which additionally uses

[3] To whom any correspondence should be addressed

the data from direct observation of the divertor. The method uses (i) semi-analytical model [7] for the spectral line shape of the Balmer-alpha line of hydrogen isotopes, taking into account the essential asymmetry of the spectral line shape caused by the fact that fast atoms dominate in the flow from the wall into plasma (see Ballistic Model [8,9]), and (ii) the model [10] of the recovery of basic parameters (namely, effective temperatures and relative contributions) of non-Maxwellian fractions of the velocity distribution function (VDF) of neutral hydrogen in the SOL. The method involves the following successive procedures: (i) recovery of the spatial distribution of the temperature in the divertor, using the spectra obtained via direct observation of the divertor; (ii) calculation of the DSL spectrum from the recovered distribution of temperature; (iii) solving a multi-parametric inverse problem, taking into account the contributions of the sections on the high magnetic field side (HFS) and low magnetic field side (LFS) of the observation chord in the SOL in the main vacuum chamber, and the contribution of the DSL. The results [11, 12] of the first applications of the developed model to interpret the data from the JET-ILW tokamak experiments, which simulate the conditions of occurrence of the DSL in ITER, are discussed.

2. Basic equations

A typical geometry of complex diagnostics, which includes simultaneous measurements of the emission of neutral hydrogen in the SOL plasma in the main chamber and in the divetor is shown in Figure 1 for ITER. The 2D profile of emissivity, calculated by S.W. Lisgo in the frame of the approach [3-6] for the flat-top of the so-called inductive discharge (the fusion gain Q=10) in a scenario with low density in the far SOL, in the L-mode regime (scenario "d"), is shown.

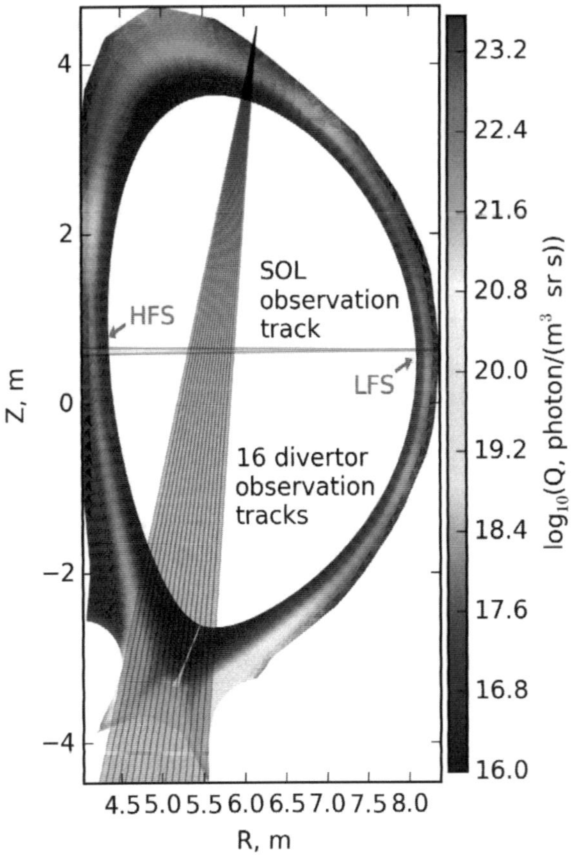

Figure 1. The 2D distribution of the Balmer-alpha emissivity in the SOL and divertor in ITER, in logarithmic scale. Possible layouts of the radial horizontal line of sight (track) from the equatorial port plug and the 16 tracks for viewing from the top port-plug down towards the divertor are shown. The HFS and LFS sections of the observation chord in the SOL in the main chamber are shown.

The Balmer-alpha spectral line shapes of hydrogen isotopes in divertor are symmetric, whereas the line shapes for emission from the SOL is asymmetric and should be described by the model [7]. Here for simplicity we assume that the σ-components of the Zeeman triplet are filtered out using the polarizers (this is the case for current experiments on JET-ILW tokamak), extension to allowance for the entire Zeeman triplet does not change the principles of the method. The inverse problem of recovering the characteristic temperature of the neutral atoms in the divertor is formulated as follows:

$$\sum_{j=1}^{N_{tr}}\left(\tilde{S}_{tr}^{\exp}(\lambda_j)-\left\{\sum_{i=1}^{M}x_{tr}^{(i)}\left(\begin{array}{c}(1-C_{H,tr})\tilde{F}_{Gauss}\left(\lambda_j-\lambda_{D\alpha},T_{tr}^{(i)}\right)\\ +C_{H,tr}\tilde{F}_{Gauss}\left(\lambda_j-\lambda_{H\alpha},T_{tr}^{(i)}\right)\end{array}\right)\right\}\right)^2 \xrightarrow[\mathbf{x}_{tr},\mathbf{T}_{tr},C_{H,tr}]{} \min,\ tr=1:Tr_{\max}^{div}, \qquad (1)$$

where λ_j is the wavelength that corresponds to the spectral channel (pixel) with number j; N_{tr} is the total number of pixels in the Balmer-alpha spectral interval chosen for the interpretation for the track with number tr; $\tilde{S}_{tr}^{\exp}$ is the normalized line shape of the spectrum, measured at the current time interval on the track tr; $\lambda_{D\alpha}$ and $\lambda_{H\alpha}$ are the wavelengths of the Balmer-alpha line center of the deuterium and hydrogen, respectively; $\tilde{F}_{Gauss}$ is the normalized Gaussian function; i is the index of the fraction of atoms with the certain temperature; M is the total number of those fractions the temperature of which should be recovered (it is possible to recover two or three fractions); $Tr_{\max}^{div}$ is the total number of tracks for direct observation of the divertor; $T_{tr}^{(i)}$ and $x_{tr}^{(i)}$ are, respectively, the temperature and partial contribution (statistical weight) of the i–th fraction of atoms on the track tr; $C_{H,tr}$ is the partial contribution of the hydrogen (H) to the total (i.e. H + D) intensity observed on the track tr. Here we assume the equality of temperatures of hydrogen and deuterium atoms.

Using the recovered temperature distribution in the divertor we may predict the normalized spectrum (i.e. the line shape) of the DSL as follows:

$$\tilde{S}_{DSL}(\lambda)=\sum_{tr=1}^{Tr_{\max}^{div}}\frac{S_{tr}R_{tr}}{\sum_{tr'=1}^{Tr_{\max}^{div}}S_{tr'}R_{tr'}}\sum_{i=1}^{M}\hat{x}_{tr}^{(i)}\left(\begin{array}{l}(1-\hat{C}_{H,tr})\left(\begin{array}{c}C_{\pi}^{(DSL)}\tilde{F}_{Gauss}\left(\lambda_j-\lambda_{D\alpha},\hat{T}_{tr}^{(i)}\right)\\ +\dfrac{\left(1-C_{\pi}^{(DSL)}\right)}{2}\left(\begin{array}{l}\tilde{F}_{Gauss}\left(\lambda_j-\lambda_{D\alpha}-\Delta\lambda_{D,tr}^{Zeem}(R_{tr}),\hat{T}_{tr}^{(i)}\right)\\ +\tilde{F}_{Gauss}\left(\lambda_j-\lambda_{D\alpha}+\Delta\lambda_{D,tr}^{Zeem}(R_{tr}),\hat{T}_{tr}^{(i)}\right)\end{array}\right)\end{array}\right)\\ +\hat{C}_{H,tr}\left(\begin{array}{c}C_{\pi}^{(DSL)}\tilde{F}_{Gauss}\left(\lambda_j-\lambda_{H\alpha},\hat{T}_{tr}^{(i)}\right)\\ +\dfrac{\left(1-C_{\pi}^{(DSL)}\right)}{2}\left(\begin{array}{l}\tilde{F}_{Gauss}\left(\lambda_j-\lambda_{H\alpha}-\Delta\lambda_{H,tr}^{Zeem}(R_{tr}),\hat{T}_{tr}^{(i)}\right)\\ +\tilde{F}_{Gauss}\left(\lambda_j-\lambda_{H\alpha}+\Delta\lambda_{H,tr}^{Zeem}(R_{tr}),\hat{T}_{tr}^{(i)}\right)\end{array}\right)\end{array}\right)\end{array}\right), \qquad (2)$$

where S_{tr} is the wavelength-integrated intensity on the track tr; R_{tr} is the major radius of the point of the maximum emissivity on the track tr (the $S_{tr}\cdot R_{tr}$ product recalculates the observation volume of the track tr to the volume of the respective emitting toroidal ring); $\hat{x}_{tr}^{(i)}$, $\hat{T}_{tr}^{(i)}$ and $\hat{C}_{H,tr}$ are the input parameters found by solving the problem (1); $C_{\pi}^{(DSL)}$ is the partial contribution of the Zeeman π-component to the total line shape of the DSL, the output parameter to be recovered from Equations (3)-(6)).

The inverse problem of interpreting the measurements in the main chamber is formulated as follows:

$$\sum_{j=1}^{N_{tr}}\left(\begin{array}{c}\tilde{S}_{tr}^{\exp}(\lambda_j)-x_{tr}^{DSL}\tilde{S}_{DSL}\left(C_{\pi,tr}^{(DSL)},\lambda_j\right)\\ -\sum_{p=1}^{2}\left\{\begin{array}{c}\sum_{i=1}^{M}x_{tr}^{(i)}\left(\begin{array}{c}(1-C_{H,tr})\tilde{F}_{Maxw}^{SOL}(\lambda_j,\Delta\lambda_{D,tr,p}^{Zeem},\lambda_{D\alpha},C_{\pi,tr,p}^{(SOL)},T_{tr,p}^{(i)})\\ +C_{H,tr}\tilde{F}_{Maxw}^{SOL}(\lambda_j,\Delta\lambda_{H,tr,p}^{Zeem},\lambda_{H\alpha},C_{\pi,tr,p}^{(SOL)},T_{tr,p}^{(i)})\end{array}\right)\\ +\sum_{i=2}^{M}x_{tr}^{(M+i)}\left(\begin{array}{c}(1-C_{H,tr})\tilde{F}_{Non-Maxw}^{SOL}(\lambda_j,\Delta\lambda_{D,tr,p}^{Zeem},\lambda_{D\alpha},C_{\pi,tr,p}^{(SOL)},T_{tr,p}^{(i)},\Lambda_{tr,p}^{(i)})\\ +C_{H,tr}\tilde{F}_{Non-Maxw}^{SOL}(\lambda_j,\Delta\lambda_{H,tr,p}^{Zeem},\lambda_{H\alpha},C_{\pi,tr,p}^{(SOL)},T_{tr,p}^{(i)},\Lambda_{tr,p}^{(i)})\end{array}\right)\end{array}\right\}\end{array}\right)^2 \xrightarrow[\substack{C_{\pi,tr}^{(DSL)},\\ \mathbf{x}_{tr},\mathbf{T}_{tr},\\ \Lambda_{tr},C_{H,tr}}]{} \min,\quad tr=1:Tr_{\max}^{SOL}, \tag{3}$$

$$F_{Maxw}^{SOL}(\lambda,\Delta\lambda_D^{Zeem},\lambda_{D\alpha},C_\pi,T)=\left(\begin{array}{c}C_\pi F_{Gauss}(\lambda-\lambda_{D\alpha},T)\\ +\frac{1-C_\pi}{2}\left(F_{Gauss}\left(\lambda+\Delta\lambda_D^{Zeem}-\lambda_{D\alpha},T\right)+F_{Gauss}\left(\lambda-\Delta\lambda_D^{Zeem}-\lambda_{D\alpha},T\right)\right)\end{array}\right), \tag{4}$$

$$F_{Non-Maxw}^{SOL}(\lambda,\Delta\lambda_D^{Zeem},\lambda_{D\alpha},C_\pi,T,\Lambda)=\left(\begin{array}{c}C_\pi F_{Asym}(\lambda-\lambda_{D\alpha},T,\Lambda)\\ +\frac{1-C_\pi}{2}\left(F_{Asym}\left(\lambda+\Delta\lambda_D^{Zeem}-\lambda_{D\alpha},T,\Lambda\right)+F_{Asym}\left(\lambda-\Delta\lambda_\alpha^{Zeem}-\lambda_{D\alpha},T,\Lambda\right)\right)\end{array}\right), \tag{5}$$

$$F_{Asym}(\Delta\lambda,T,\Lambda)=F_{Gauss}(\Delta\lambda,T)\exp\left(-\Lambda/|\Delta\lambda|\right)\eta\left(-\Delta\lambda(\mathbf{k},\mathbf{l})\right), \tag{6}$$

where p indicates the emission section (either HFS or LFS) of the SOL; summations over i go over Maxwellian and non-Maxwellian fractions of atoms; the second sum starts with $i = 2$ because of the lack of a «cold» non-Maxwellian fraction (the temperatures of the «warm» Maxwellian and «warm» non-Maxwellian fractions are assumed to be equal); x_{tr}^{DSL} is the fraction of the DSL in the total signal and should be recovered (it varies in the range of about 10% around the universal value which is a function of the divertor emissivity, geometry of the main chamber and wall reflectivity; this value is to be found for the current intensity of divertor emission by processing the observation data from many discharges); $\Lambda_{tr,p}^{(i)}$ is the characteristic wavelength shift for the spectral contribution of the i-th fraction of non-Maxwellian atoms, which describes the attenuation of the inward flux of atoms on the track tr (see [7] for the details); $\eta(x)$ is the Heaviside function; **k** is the direction of atomic flux from the wall to plasma in the given section of the SOL; **l** is the direction from the detector to the observation point.

The results of the first applications of the developed model to interpret the data from the JET-ILW tokamak are presented in [11, 12]. Typical results for the D-alpha spectral line on the divertor stage of the discharge were presented for the radial and tangential tracks from the equatorial port. The recovered temperatures and the fractions of Maxwellian and non-Maxwellian atoms have been found for HFS and LFS sections of the SOL. The recovered time behavior of the fractions of different sources of the signal (HFS and LFS sections of the SOL, and the DSL) on these tracks indicated a significant fraction of the DSL (up to 50%) in the total signal throughout major part of the discharge in the JET-ILW tokamak. Extrapolation to ITER has suggested a stronger effect of the DSL in ITER.

3. Conclusions

A method is developed (and implemented on JET-ILW) for interpreting the data from the Balmer-alpha high-resolution spectroscopy in the edge plasma with account of the divertor stray light (DSL). The method uses, additionally to conventional H-alpha diagnostics in the main chamber, the data from direct observation of the divertor. The method allows to (i) separate, in the total signal, the contributions from three sources, the HFS and LFS sections on the observation chord in the main chamber, and the DSL, (ii) recover the temperature of the hydrogen and deuterium atoms, (iii) evaluate the isotope ratio in a H+D mixture.

Acknowledgements

The authors are grateful to S.W. Lisgo, A.S. Kukushkin, V.S. Lisitsa, S. Brezinsek, A.V. Gorshkov, M. von Hellermann, M.B. Kadomtsev, V. Kotov, M.G. Levashova, V.A. Shurygin, M.F. Stamp, E. Veshchev, D.K. Vukolov, K.Yu. Vukolov for their collaboration in studies on the ITER Main Chamber H-alpha (and Visible Light) Spectroscopy Diagnostics.

References

[1] Kukushkin A B *et al.* 2012 *Proc. 24th IAEA FEC (San Diego, USA)* ITR/P5-44
[2] Kajita S *et al.* 2013 *Plasma Phys. Contr. Fusion* **55** 085020 (9 pp)
[3] Kukushkin A S *et al.* 2009 *Nucl. Fusion* **49** 075008
[4] Braams B J 1986 *PhD thesis* (Utrecht: Rijksuniversitet)
[5] Reiter D, Baelmans M and Boerner P 2005 *Fusion Sci. Tech.* **47** 172
[6] Lisgo S W *et al.* 2011 *J. Nucl. Mater.* **415** S965
[7] Kukushkin A B *et al.* 2014 JPCS **548** 012012 (5 pp)
[8] Lisitsa V S *et al.* 2014 *Atoms* **2** 195-206
[9] Kadomtsev M B *et al.* 2012 *Proc. 39th EPS Conf. (Stockholm, Sweden)* P4.093
[10] Neverov V S *et al.* 2015 *Plasma Phys. Rep.*, **41** 103-111
[11] Kukushkin A B *et al.* 2014 *AIP Conf. Proc.* **1612** 97-100
[12] Kukushkin A B *et al.* 2014 *Proc. 25th IAEA FEC (St. Petersburg, Russia*) EX/P5-20 (*JET Preprint* EFDA–JET–CP(14)06/33)

On resolving the optical spectra of the edge plasma radiation against a strong background of the divertor stray light

L I Ognev[1,3] **and V S Lisitsa**[1,2]
[1]National Research Centre "Kurchatov Institute", Moscow 123182 Russia
[2]National Research Nuclear University "MEPhI", Moscow, Russia

E-mail: liognev@gmail.com

Abstract. A method of detecting a weak radiation of the Balmer lines from the hydrogen isotopes in the edge plasma of a tokamak (SOL) on the strong optical background formed by reflection of the light emitted from the divertor plasma is proposed. The method is based on a parallel registration of the radiation from the optical dump and from the wall near the dump. Calibration of the attenuation of the stray light radiation by the optical dump is being performed in the course of the measurements, using the intensity of the high-frequency pulsations that are not correlated with the edge plasma emission. Statistical estimates show that a 20-fold attenuation of the light by the dump allows registration of the weak edge plasma radiation making down to 0.5% of the background intensity.

1. Introduction

The issue of registering the Balmer line radiation from the hydrogen isotopes in the edge tokamak plasma (SOL) has become particularly important after first assessments of the background formed by reflection of the powerful light radiation from the divertor plasma. The ratio of the divertor to the SOL radiation intensity in the H-alpha line in modern tokamaks (JET) is about an order of magnitude. Given the parameters of ITER, this ratio can reach a value of 10^2 [1].

2. Methods of detecting a weak radiation on a strong background

Several methods have been proposed to solve this problem [1, 2]. Utilisation of an optical light dump (an absorbing cavity in the wall) was proposed in [1]. The method [2] is based on a solution of the inverse problem of recovering the SOL signal using the data of measurements along different chords, including those viewing the divertor. Parallel detection of the radiation from the dump and wall ("double chords" scheme) improves the stability of the method.

2.1. The dump calibration method using high-intensity fluctuations of the background

In this paper we analyze the possibility of detecting a weak SOL signal on the background of a strong divertor radiation in the scheme with the light dump [2] ("double chords"), taking advantage of the statistical independence of the fluctuations of the signals from both sources. The typical frequencies of

[3] To whom any correspondence should be addressed

such fluctuations in modern tokamaks, according to the observations of the spectra of optical radiation from helium [3] and hydrogen [4], reach the values up to 100 kHz. In the method proposed, the high frequency part of the fluctuation spectrum is used to calibrate the optical dump.

This method is based on comparing the intensity of the radiation beams reflected from the attenuating dump I_2 and from the first wall without attenuation I_1 (figure 1). Both beams include, as a minor fraction, the radiation from the neutral atoms in the edge plasma (SOL), x_1 and x_2, respectively, which is registered without attenuation and is the useful signal to be resolved. The background radiation reflected from the vicinity of the dump, y_1, is not attenuated – unlike that reflected from the dump itself, y_2 (figure 2). Therefore, $x_1 \approx x_2 = x << y_1$, $y_2 = \alpha y_1$, $x_1 + y_1 = I_1(t)$, $x_2 + y_2 = I_2(t)$, where α is the light attenuation factor in the dump. Now the intensity of the weak signal can be found as:

$$x = (I_2 - \alpha I_1)/(1-\alpha) \tag{1}$$

Diagnostic requirements for monitoring the H-alpha signal foresee the time resolution of 0.1 s. Therefore, the time-average values should be taken for the intensities I_1 and I_2 in equition 1. The intensity of the high frequency part of the fluctuation spectrum in the text below will be noted as $\tilde{I}$. So, in order to detect the weak signal on the strong background one needs to know the reflectivity of the light beam dump, which may vary in time over the measurement period of the signal $T_1 = 0.1$ s. Unlike the "double chords" [2], this method does not require detailed information about the shape of the line.

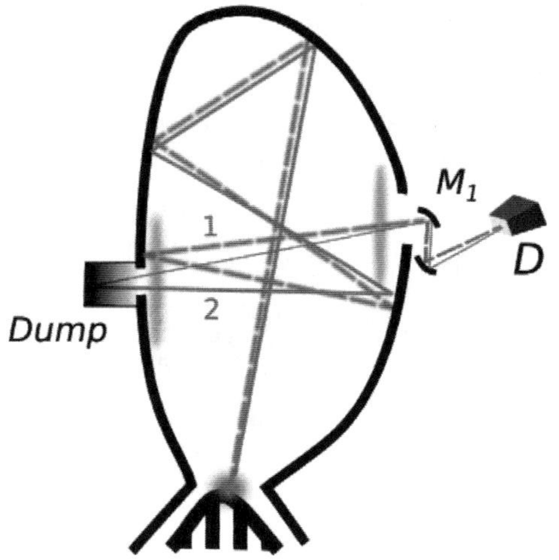

Figure 1. Scheme of the formation of the two-channel signal I_1 and I_2 by the reflection of the divertor stray light from the wall (chord 1) and from the optical dump (chord 2) with the addition of the radiation contribution of the edge plasma. M_1 is the first diagnostic mirror, D the detector with separate registration of the channels. The radiation zones are shadowed.

The reflectivity of the light dump can be calculated from the time-averaged power of the fluctuations of the high intensity divertor radiation in the light beams reflected from the dump and from the chamber wall as

$$\alpha(t) = \sqrt{\langle \tilde{y}_2^2 \rangle} \Big/ \sqrt{\langle \tilde{y}_1^2 \rangle} \approx \sqrt{\langle \tilde{I}_2^2 \rangle} \Big/ \sqrt{\langle \tilde{I}_1^2 \rangle} \tag{2}$$

since, due to the statistical independence, $\langle \tilde{I}_i^2 \rangle = \langle \tilde{y}_i^2 \rangle + \langle \tilde{x}_i^2 \rangle$ and $\langle \tilde{x}_i^2 \rangle << \langle \tilde{y}_i^2 \rangle$.

2.1.1 Statistical accuracy of the dump effectiveness and the bandwidth of the background fluctuations. High intensity fluctuations of the H-alpha line are collected in a frequency band [ω_2, ω_3] and subjected to quadratic detecting followed by isolation of the low-frequency signal components $\langle \tilde{I}_i^2 \rangle$ in the diagnostic range [0, ω_1]. We will use the relation between the dispersion of the current through the quadratic detector squared and the frequency ranges used for registration of the high- and low-frequency signals in the model of the white noise with the Gaussian statistics [5]:

$$\frac{<(\delta<\tilde{I}_i^2>)^2>}{(<\tilde{I}_i^2>)^2} = \int_0^{\omega_1} S_f(\omega)d\omega \Big/ \int_0^{\omega_1} S_0(\omega)d\omega = \frac{4(<\tilde{I}_2^2>)^2 \frac{\pi}{\Delta\omega}\cdot\omega_1}{(<\tilde{I}_i^2>)^2} = \frac{4\pi\omega_1}{\omega_3-\omega_2} \tag{3}$$

where $S_f(\omega)$ and $S_0(\omega)$ are the spectral densities of the fluctuations of the processed signal and its constant component. Since $\delta\alpha/\alpha \sim (1/2)\cdot\delta(\langle\tilde{I}_i^2\rangle)/\langle\tilde{I}_i^2\rangle$ from (3) one obtains $\delta\alpha/\alpha \sim (1/2)\cdot\sqrt{4\pi\omega_1/(\omega_3-\omega_2)}$

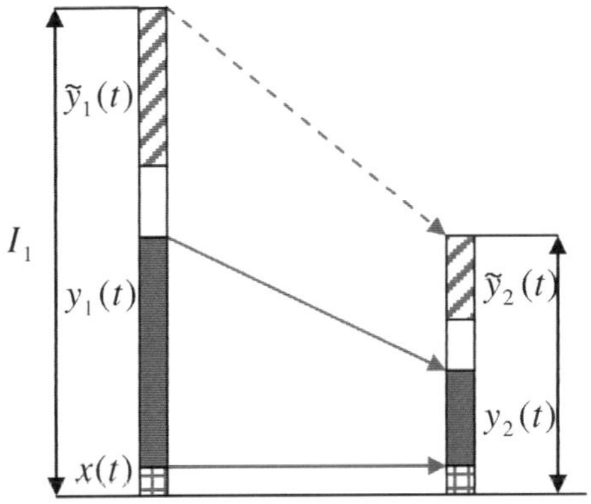

Figure 2. Schematics of the measurement of the optical radiation signal from the neutral hydrogen atoms (*x (t)*) on the background of the divertor stray light (*y (t)*). The total radiation intensity *I_1 (t)* and the total intensity of the radiation from the optical dump *I_2 (t)* is what is actually measured. The shaded areas correspond to the rapidly oscillating component of the radiation coming from the divertor.

The bandwidth of the high-frequency filter necessary to satisfy the accuracy $\delta\alpha/\alpha \le 0.1$ can be found from the relation $4\pi\omega_1/0.04 \le \omega_3 - 10\omega_1$. Hence $\omega_3 > (4\pi/0.04+10)\omega_1$, and for $T = 0.1$ s, the frequency of the high-frequency pulsations used for the dump calibration $f = 3200$Hz is adequate for the dump calibration. This corresponds to the minimum reference time interval for the noise $\Delta t = 0.3\cdot10^{-3}s$. The required photon flux through the high-frequency filter in the interval of $\Delta t/2$ is $N/(\Delta t/2) = 100/(0.3\cdot10^{-3}s/2) = 7\cdot10^5$ph/s, and this can indeed be registered in experiment.

2.1.2 Sensitivity of the method. The uncertainty of the $\bar{x}(t)$ calculation is attributed to variations in the reflectivity of the dump $\alpha(t) = \bar{\alpha}(t) + \delta\alpha$. Let us express the variation through the partial derivative of $x = a(\alpha, I_2)$ with respect to α: $\delta x = (\partial x/\partial\alpha)\cdot\delta\alpha$. After substitution $I_2 = \alpha(t)[I_1(t) - x] + x$, taking into account the smallness of $x \ll I_1$ we obtain

$$\sqrt{\langle(\delta x)^2\rangle} \approx I_1\sqrt{\langle(\delta\alpha)^2\rangle}\Big/(1-\alpha) \tag{4}$$

Therefore, minimization of $\delta\alpha$, rather than the smallness of α itself, is important for improving the sensitivity. Assuming $\delta\alpha/\alpha \le 0.1$, if $\alpha \le 0.05$ [1], the minimum detectable signal-to-noise ratio x_{min}/I_1 can be estimated as less than 0.005.

3. Acknowledgements

The authors wish to thank A. B. Kukushkin and V. S. Neverov for fruitful discussions. This work was supported in part by the grant NSH-3328.2014.2 of the President of the Russian Federation for the state support of leading scientific schools of the Russian Federation.

References

[1] Andreenko E N, Alekseyev A G and Gorshkov A V 2014 *AIP Conference Proceedings* **1612** 171
[2] Kukushkin A B et al 2012 *Book of Abstracts, 24th IAEA Fusion Energy Conference, San Diego, USA, 8-13 October 2012* ITR/P5-44
[3] Rosmej F B, Ohno N, Nakamura S and Kajita S 2008 *Contrib. Plasma Phys.* **48** 243
[4] Zielinski J J, van der Meiden H J, Morgan T W, Hoen M H J, Schram D C and De Temmerman G 2014 *Appl. Phys. Lett.* **104** 124102
[5] Akhmanov S A, Dyakov Yu E and Chirkin A S 1981 *Introduction to statistical radiophysics and optics* (Moscow: Nauka)

Analysis of measurement errors for Thomson diagnostics of non-Maxwellian plasmas in tokamak reactors

P A Sdvizhenskii[1,5], A B Kukushkin[1,2], G S Kurskiev[3], E E Mukhin[3] and M Bassan[4]

[1] National Research Center "Kurchatov Institute", Kurchaton square 1, Moscow, 123182, Russian Federation

[2] National Research Nuclear University MEPhI, Kashirskoe shosse 31, Moscow, 115409, Russian Federation

[3] Ioffe Physical Technical Institute, Politekhnicheskaya 26, St. Petersburg, 194021, Russian Federation

[4] ITER Organization, Route de Vinon sur Verdon, 13115 St Paul Lez Durance, France

E-mail: sdvinpt@gmail.com

Abstract. The study is stimulated by the expected noticeable deviation of the electron velocity distribution function (eVDF) from a Maxwellian under condition of a strong auxiliary heating of electron plasmas in tokamak-reactors. The key principles of accuracy estimation of the Thomson scattering diagnostic of non-Maxwellian plasmas in tokamak-reactors are presented. The algorithm extends the conventional approach to the assessment of non-Maxwellian plasmas measurements errors for a broad class of deviations of the eVDF from a Maxwellian. The algorithm is based on solving the inverse problem many times to determine main parameters of the eVDF with allowance for all possible sources of error and statistical variation of the input parameters of the problem. The method is applied to a preliminary analysis of the advantages of the formerly suggested use of various wavelengths of probing laser radiation in the Thomson diagnostics of non-Maxwellian plasma on the example of the core plasma Thomson scattering diagnostic system which is under design for ITER tokamak. The results obtained confirm the relevance of the diversification of the probing laser radiation wavelength.

1. Introduction

The Thomson Scattering (TS) diagnostics of high electron temperature T_e in the core plasma in tokamak reactors (DEMO and ITER) has to operate in a limited spectral range of a strongly broadened TS spectrum and allow for possible deviation of the electron velocity distribution function (eVDF) from a Maxwellian under condition of a strong auxiliary heating. The analysis [1] of opportunities to enhance the accuracy of the core plasma TS diagnostics suggested the use of several probing wavelengths to increase the number of signals from different spectral channels and to be able to interpret the TS data with allowance for possible deviation from a Maxwellian eVDF in the range of weakly/moderate superthermal energy, with account of data for higher energy from other diagnostics (e.g., from that suggested in [2]).

In this paper we present the results of further development of the method [3-5] for detailed accuracy analysis of the Thomson scattering diagnostics of non-Maxwellian plasmas in tokamak-

[5] To whom any correspondence should be addressed

reactors. The method is based on the correct error assessment to determine main parameters of the eVDF from the TS spectrum measurements with allowance for all possible sources of error. The method is applied for analyzing the advantages of using several probing wavelengths for the core plasma TS diagnostics for Maxwellian and non-Maxwellian plasmas via solving an inverse problem for error assessment. This includes evaluation of the possibility to recover, from visible and infrared light spectral range ~400—1000 nm, (i) the electron temperature T_e of the bulk (Maxwellian) plasma up to ~ 40 keV and (ii) a moderate anisotropy of the eVDF in electron pitch angles in the weakly/moderate superthermal energy range. A particular example of the ITER core plasma TS system design is considered for a test of suggested approach, and a comparative analysis of conservative and advanced approaches is given.

2. Algorithm of error assessment

The algorithm to assess the measurement error should be formulated in the frame of a synthetic diagnostics. Such a diagnostic generates "phantom" experimental data and allows direct comparison of the pristine (i.e. taken as known, "assumed") and the recovered values of diagnosed parameters. The number of photoelectrons in a given spectral channel due to Thomson scattering is as follows:

$$\left[N^{(j)}_{\text{ph-el}}\right]_{\text{Las}} = \frac{E_{\text{Las}}}{\hbar\omega_0}\Delta x_{\text{Las}}\,\Delta\Omega_{\text{pupil}}\,n_e r_0^2 \int\limits_{\Delta X_j} \eta_q(X)\eta_{\text{tr}}\,\eta_{\text{fib}}(X)\sigma_{\text{TS}}\,dX\,, \tag{1}$$

where $X = \lambda/\lambda_0 - 1$, ΔX_j is the spectral width of the j-th channel; λ and λ_0, the wavelength of scattered and probing radiation; E_{Las} , laser pulse energy; Δx_{Las}, the length of a radiating cylinder viewed by the detector ("scattering length"); n_e, electron density; $\Delta\Omega_{\text{pupil}} = \Delta S_{\text{pupil}}/(\mathbf{r}_{\text{emis}} - \mathbf{r}_{\text{pupil}})^2$, where ΔS_{pupil} is the area of the pupil, and the vectors stand for the coordinate of the radiating cylinder and the pupil; r_0, electron classical radius; $\mathbf{n}_0$ and $\mathbf{n}$, directions of incident and scattered light; $\mathbf{e}_0$ and $\mathbf{e}$, polarization vectors of incident and scattered light; the factors η allow for the properties of optical system. The normalized cross-section σ_{TS} of the Thomson scattering (i.e. incoherent, non-collective scattering by relativistic electrons with neglect of the electron's recoil in the elementary act of scattering the photon) can be calculated with formulas [6-8].

In the present paper we consider the following model eVDF, which accords with results of the eVDF numerical modelling studies:

$$f(\mathbf{p}) = (1-\delta_{\text{Hot}})\cdot f_{\text{Maxw}}(\mathbf{p}) + \delta_{\text{Hot}}\cdot f_{\text{Hot}}(\mathbf{p}_\parallel,\mathbf{p}_\perp)\,, \tag{2}$$

$$f_{\text{Hot}}(\mathbf{p}_\parallel,\mathbf{p}_\perp) = C_{\text{Hot}}\exp\left[-mc^2\left(\frac{\sqrt{1+(p/mc)^2}-1}{p^2}\right)\left(\frac{p_\parallel^2}{T_{e\parallel}}+\frac{p_\perp^2}{T_{e\perp}}\right)\right] \tag{3}$$

Here $f_{\text{Maxw}}(\mathbf{p})$ is the relativistic Maxwellian distribution; C_{Hot}, normalization factor (in momentum space); $\mathbf{p}_\parallel$ and $\mathbf{p}_\perp$, the components of momentum $\mathbf{p}$, parallel and perpendicular to local magnetic field; δ_{Hot}, total fraction of superthermal electrons. The background signal is assumed to be measured in a separate time window of similar duration. Different lasers are assumed to act in different but very close time instants.

The TS-based eVDF diagnostic algorithm [3-5] includes the calculation of the number of photoelectrons, $[N^{(j)}_{\text{ph-el}}]_{\text{Laser}}$, in a given spectral channel of the detector (where j numerates spectral channels) in terms of the normalized cross-section σ_{TS} of the Thomson scattering, which is averaged over the assumed model eVDF (2), (3). For each spectral channel, the difference between total, $[N^{(j)}_{\text{ph-el}}]_{\text{Total}}$, and background, $[N^{(j)}_{\text{ph-el}}]_{\text{Backgr}}$, signals is calculated. Randomization of input parameters is applied to estimate the diagnostic's accuracy (for $[N^{(j)}_{\text{ph-el}}]_{\text{Laser}} \gg 1$, one can take a Gaussian with the average value $[N^{(j)}_{\text{ph-el}}]_{\text{Laser}}$ and the mean square deviation $\sigma_j = \left\{\left[N^{(j)}_{\text{ph-el}}\right]_{\text{Las}} + 2\left[N^{(j)}_{\text{ph-el}}\right]_{\text{Backgr}} + 2\left[\sigma_j^2\right]_{\text{Ampl}}\right\}^{1/2}$, where $\left[\sigma_j\right]_{Ampl}$ describes the noise of

Journal of Physics: Conference Series **666** (2016) 012006 doi:10.1088/1742-6596/666/1/012006

amplifier). The values of the sought-for parameters $\xi = \{T_e, T_{e\parallel}, T_{e\perp}, n_e, \delta_{Hot}\}$ are recovered via solving the inverse problem, which relies on the minimization of the difference between the "phantom" experimental laser-scattering signal and the respective variable calculated signal:

$$\sum_{\lambda_0}\sum_{j}\left|\left\{\left[N^{(j)}_{\text{ph-el}}(\xi_{\text{assum}})\right]_{\text{Total}} - \left[N^{(j)}_{\text{ph-el}}\right]_{\text{Backgr}}\right\}_{\text{Random}} - \left[N^{(j)}_{\text{ph-el}}(\xi)\right]_{\text{Las}}\right| \xrightarrow[\xi]{} \min, \quad (4)$$

where the summation goes over all lasers and spectral channels. The error of recovery of the value of parameter B, denoted as Err, is defined as an average of the normalized mean square deviation from the assumed value of B over statistical distribution of the recovered values of B:

$$\text{Err}(B^{\text{assumed}}) \equiv 2,5\left[\left\langle\left(B/B^{\text{assumed}} - 1\right)^2\right\rangle\right]^{1/2}. \quad (5)$$

The factor 2.5 in (5) corresponds to 98% probability to find the value of B in the range $\{B^{\text{assumed}} - 2{,}5\delta B,\ B^{\text{assumed}} + 2{,}5\delta B\}$, where $\delta(B) = [\langle(B - B^{\text{assumed}})^2\rangle]^{1/2}$. For a non-Maxwellian eVDF, one has to analyze the accuracy of measuring the mean kinetic energy:

$$\langle E_{\text{kin}}\rangle = T_{\text{эф}} - 1 - K_2'(1/T_{\text{эф}})/K_2(1/T_{\text{эф}}); \quad E_{\text{kin}} = \sqrt{(p/mc)^2 + 1} - 1, \quad (6)$$

where K_2 is the MacDonald function of the 2nd order, the prime denotes the derivative. In the non-relativistic case, this yields the well-known relation $\langle E_{\text{kin}}\rangle = (3/2)T$.

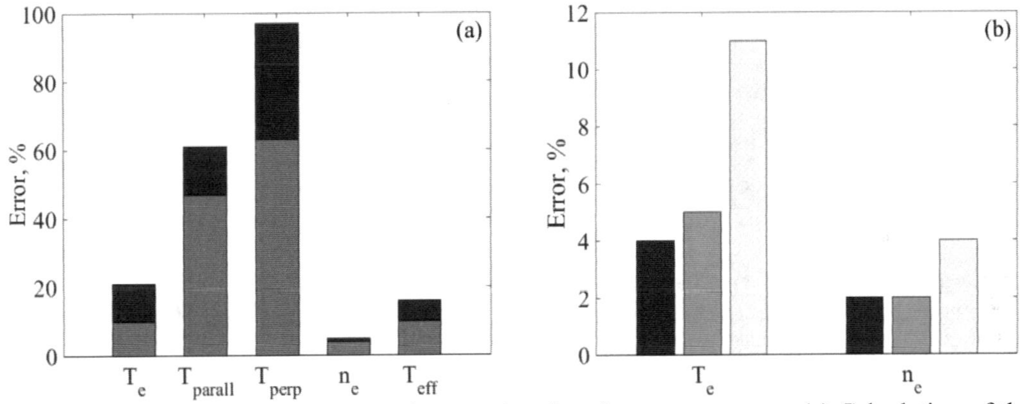

Figure 1. The comparison of errors of recovering the plasma parameters. (a) Calculation of the phantom spectrum for the non-Maxwellian eVDF and interpretation assuming the non-Maxwellian eVDF of Equations (2), (3): conventional approach with one color probing radiation 1064 nm (blue); multicolor (946; 1064; 1320 nm) probing radiation (red). (b) Calculation of the phantom spectrum for the non-Maxwellian eVDF and interpretation assuming a Maxwellian eVDF: one color probing radiation 1064 nm (blue); the same but for multicolor (1064; 1320 нм) probing radiation (green). Calculation of the phantom spectrum for the Maxwellian eVDF and interpretation assuming the non-Maxwellian eVDF of Equations (2), (3) (yellow).

Figure 1 shows the results for 200 runs of solving the inverse problem for the following input parameters which are considered in the ongoing analysis of the core plasma TS system in ITER. Figure 1(a) shows the results in the case of calculating the phantom spectrum for a non-Maxwellian eVDF and interpreting it assuming the non-Maxwellian eVDF of Equations (2), (3): the case of three lasers with λ_0 = 1064 nm and the case of three lasers with different probing wavelengths λ_0 = 946; 1064; 1320 nm; $n_e = 3\times10^{19}$ m^{-3}; E_{Las} = J (the same for all lasers); $\eta_{tr} = 0.168$ is the total transmission factor of optical system (a flat efficiency of 30% for the collection optics, a 70% packing efficiency for the fiber bundles, a 80% optical efficiency of the spectrometers); the quantum detector sensitivity, η_q, is taken to have the APDs spectral sensitivity like the Excelitas C30956EH in the range above 650 nm and like the Hamamatsu S8664 in the range below 650 nm; $[\sigma_j]_{\text{Ampl}}$ = 50; $\Delta x_{\text{Las}} = 0.07$ m;

$\Delta\Omega_{pupil} = 1.38\times10^{-3}$ sr. The angle between $\mathbf{n}_0$ and $\mathbf{n}$ is taken 159.5° that corresponds to observation of the plasma column center; linear polarization vectors $\mathbf{e}_0$ and $\mathbf{e}$ coincide and are perpendicular to scattering plane. The background signal is formed by the direct signal from the plasma Bremsstrahlung along the line of sight and the reflected-from-the-wall light which is determined by the line radiation emitted in the divertor (the background signal data are provided in [9]). Errors for the parameter δ_{Hot} are about several hundred percent (not shown in the figure 1(a)). Note that the high error of recovering the parameters, which describe particular type of the deviation from a Maxwellian (especially, parameter δ_{Hot}), does not influence the accuracy of recovering the mean energy of the non-Maxwellian eVDF. Thus, the inverse problem solution is stable with respect to recovery of the mean energy regardless of particular form of the deviation from a Maxwellian in the thermal and weakly superthermal range of electron energy.

Figure 1(b) shows the results of similar error estimations for the case where the phantom spectrum is taken for a non-Maxwellian VDF whereas the interpretation of this spectrum assumes a Maxwellian VDF, and the opposite case of a small deviation from a Maxwellian VDF and interpretation of data assuming a non-Maxwellian VDF (2), (3). Note that the increase of the errors of multicolor probing wavelength compared to the one color probing wavelength is attributed to a partial loss of accuracy in solving the inverse problem with an increased number of unknown, sought-for parameters.

3. Conclusions

The results of analysing the opportunity to enhance the accuracy of the core plasma Thomson scattering diagnostics in tokamak-reactors -- under conditions close to those in the diagnostics under design for ITER - illustrate the trend towards increasing the accuracy of measuring the parameters of non-Maxwellian electron velocity distribution function (eVDF) with increasing range of T_e measurements that confirms the relevance of the diversification of the probing laser radiation wavelength.

The method also enables one to evaluate the accuracy of measuring the mean energy of electrons in non-Maxwellian plasmas under assumption of a Maxwellian eVDF. The results obtained give quantitative illustration of the well-known high sensitivity of Thomson diagnostics to the main Maxwellian bulk and appreciably less sensitivity to a small superthermal fraction of the eVDF.

The results obtained show the necessity of a detailed analysis of the Thomson scattering diagnostics accuracy in tokamak-reactors.

The views and opinions expressed herein do not necessarily reflect those of the ITER Organization.

References

[1] Mukhin E E *et al.* 2014 *AIP Conf. Proceedings* **1612** 69-72
[2] Minashin P V and Kukushkin A B 2013 *International Review of Atomic and Molecular Physics (IRAMP)* **4** (2) 93-104
[3] Kurskiev G S, Sdvizhenskii P A, Kukushkin A B, Mukhin E E and Bassan M 2014 *Proc. 41st EPS Conf. on Plasma Phys. (Berlin, Germany)* P5.009
[4] Sdvizhenskii P A, Kukushkin A B, Kurskiev G S, Mukhin E E and Bassan M 2014 *Problems of Atomic Science and Technology, Ser. Thermonuclear Fusion (in Russian)* **37** 38—47
[5] Kurskiev G, *et al.* 2014 *Proc. 25th IAEA FEC* (S. Petersburg, Russia) EX/P5-28 (submitted to Nuclear Fusion)
[6] Kukushkin A B 1981 *Sov. J. Plasma Phys.* **7** 63
[7] Kukushkin A B and Kogan V I 1981 *Sov. J. Plasma Phys.* **7** 669
[8] Kukushkin A B, Leneva A E and Pergament V I 1983 *Problems of Atomic Science and Technology, Ser. Thermonuclear Fusion (in Russian)* **2** 88—96
[9] Kajita S *et al.* 2014 *Private communications*

Doppler reflectometry studies of plasma gradient instabilities in L-2M stellarator

N N Skvortsova[1,2,9], A Yu Chirkov[3], A A Kharchevsky[4], D V Malakhov[5,6], A K Gorshenin[7] and V Yu Korolev[8]

[1] Leading research associate, Prokhorov General Physics Institute of the Russian Academy of Sciences, Vavilova 38, 119991, Moscow, Russia

[2] Professor, National Research Nuclear University MEPhI (MEPhI), Kashirskoe shosse 31, Moscow, 115409 Russian Federation

[3] Professor, Bauman Moscow State Technical University

[4] Ph. D. Student, Moscow State Institute of Radio Engineering, Electronics and Automation

[5] Senior research associate, Prokhorov General Physics Institute of the Russian Academy of Sciences

[6] Associate professor, Bauman Moscow State Technical University

[7] Senior research associate, The Institute of Informatics Problems of the Russian Academy of Sciences

[8] Professor, Moscow State University

E-mail: mukudori@mail.ru, 89168766306@mail.ru

Abstract. Paper reports on a broadband spectrum analysis of density fluctuations in the edge plasma of the L-2M stellarator, measured using the Doppler reflectometry diagnostics. The evolution of the electron temperature-gradient (ETG) and ion temperature-gradient (ITG) instabilities in the edge plasma of the stellarator is analyzed. Two additional harmonics in the spectrum are found apart from those determined by the Doppler shift caused by poloidal rotation of plasma. The Doppler shifts of these harmonics are determined from the phase velocities of fluctuations generated by the ETG and ITG instabilities. Thus, local measurements of spectra using the Doppler reflectometry enabled us both to evaluate the poloidal rotational velocity of the plasma and describe the evolution of low-frequency plasma instabilities.

1. Introduction

It is known that plasma is a state of matter with big number of degrees of freedom so the interpretation of plasma turbulence spectra is an ill-posed problem. Studies of low-frequency plasma fluctuations in the L-2M stellarator showed that the structural turbulence is observed throughout the plasma volume. Previously it was shown that it is possible to determine the number of unstable plasma processes which form such a turbulence [1, 2].

Spectra of plasma density fluctuations measured using the Doppler reflectometry (DR) diagnostics in the edge plasma of the L-2M stellarator are broadband (the half-width exceeds maximum frequency) and correspond to structural turbulence spectra [3]. Measurement of the Doppler frequency

[9] To whom any correspondence should be addressed

shift in such spectra is difficult because the shift can depend not only on the poloidal rotation velocity of plasma but also the phase velocity of fluctuations (provided the velocities are of the same order of magnitude). The latter conditions are typical for the DR measurement in many toroidal facilities, e.g. ASDEX and TEXTOR tokamaks [4]. In stellarators, fluctuations in the spectral range measured by the DR may occur due to the gradient drift instabilities: ion temperature-gradient (ITG) instability, electron temperature-gradient (ETG) instability and the trapped electron mode (TEM) [5]. Data obtained using the DR diagnostics can be treated as local ones and, hence, comparing it to the calculated local characteristics of instabilities would be most appropriate.

In this paper the broadband spectrum measured by the DR in the L-2M stellarator was analyzed assuming that the particular random processes are responsible for formation of the measured spectra. In our study, the ITG, ETG and TEM instabilities are analyzed as contributing to the DR measurements. For all the spectra analyzed the spectral components, produced by poloidal rotation of plasma or structural turbulence modes, were extracted. The components related to poloidal plasma rotation, determined by the radial electric field, and structural turbulence modes resulted from the ITG and ETG instabilities.

2. Experimental setup and reflectometry diagnostics description

The experiments were carried out for a standard magnetic configuration of the L-2M stellarator (major radius R = 100 cm, averaged plasma radius a = 11.5 cm) when the ratio between the amplitude of fundamental harmonic of helical field and the amplitude of longitudinal field on the axis of the facility amounts to 0.228. In this case the rotational transformation angles on the magnetic axis and the boundary magnetic surface for vacuum magnetic configuration are equal to $\iota(0)/2\pi = 0{,}185$ и $\iota(a)/2\pi = 0{,}78$, respectively. The detailed description of the facility is given in [6]. The electron cyclotron (EC) heating is performed on the second harmonic of electron gyrofrequency (75.3 GHz) using a gyrotron with the power P up to 400 kW at pulse duration of 10–15 ms. Average plasma density measured along central chord was equal to $2\cdot10^{13}$ cm^{-3}. Electron temperature in the plasma core lies within the 300–1000 eV range, depending on the power of the EC heating. It should be noted that at a high heating power, the "dip" on the density profile in the plasma core occurs [7].

This effect called the "density pump-out" can be observed in tokamaks and stellarators at high power densities of external heating. In the L-2M, the "density pump-out" is distinct at the power density of EC heating of the value of $P > 2$ MW/m^3.

Long-wavelength fluctuations of plasma density ($k_\perp\rho_s \approx 0.1$–$0.2$, where ρ_s is the ion cyclotron radius) near the separatrix (r/a = 0.8–0.9) were studied by the DR diagnostics. The design of the DR diagnostics device in the L-2M stellarator, implemented in a classic two-channel scheme with Gunn diode serving as a source of probing radiation, is given in [8]. In our experiments the measurements were conducted using an optimized Doppler reflectometer. The probing radiation is injected through a quasioptical system and the signal is registered within the broad band up to 2.5 MHz [9]. The Fourier spectra measurement error (half-width and intensity) evaluated from the signals of the optimized Doppler reflectometer in the steady-state stage of plasma discharge is of 20–30%.

In the experiment the measurements were conducted at two probing frequencies of 31.2 and 34.3 GHz, at injection angles of 4°, 8°, 12°, 17° and ~ 1–2 cm away from the last closed magnetic surface. Radiation polarization corresponds to an ordinary wave in plasma. Scattering of the probing wave occurred at long-wavelength fluctuations of plasma with poloidal components of wave vector $k_\perp \approx$ 0.9–4.2 cm^{-1}.

3. Mathematical model of instabilities in the edge plasma in stellarator

The class of gradient drift instabilities includes the following modes: ITG, ETG and TEM. In classical works, the characteristic parameters of the ITG and ETG modes are defined as follows: for ITG, frequency $\omega \sim \omega_{*i}$, transverse (with respect to the magnetic field) wave number $k_\perp \sim 1/\rho_{Ti}$; for ETG, $\omega \sim \omega_{*e}$, $k_\perp \sim 1/\rho_{Te}$, where ω_{*i} and ω_{*e} are the frequencies of the diamagnetic drift of ions and electrons, ρ_{Ti} and ρ_{Te} are the cyclotron radii of ions and electrons, calculated according to their thermal velocities [13]. The real part of the frequency of the ion mode $\omega_R < 0$, for electron mode one has $\omega_R > 0$. For tokamaks the following classification is used: for ITG, $k_\perp \rho_s \lesssim 0.5$; for TEM, $k_\perp \rho_s \lesssim 2$; for ETG, $k_\perp \rho_s > 0.5$, where ρ_s is the ion cyclotron radius, calculated from the electron temperature [14]. These classifications are conditional. Depending on the conditions of the particular magnetic configurations the parameters of modes and directions of their propagation may differ significantly from the above [15]. Also there's not always a clear border between the ranges of ion and electron modes, because their overlapping is possible [5]. Within gyrokinetics approach the instability in plasma is associated with the resonant denominator in the perturbation of the distribution function:

$$\omega + \omega_{Dj} - k_\parallel v_\parallel \approx 0 \tag{1}$$

где ω is the complex value of frequency, ω_{Dj} is the orbit-averaged frequency of magnetic drift of particle of the species j ($j = i, e$), $k_\parallel$ is the longitudinal (relative to the magnetic field) component of the wave vector, $v_\parallel$ is the ongitudinal component of the particle velocity.

For the passing particles, one has $\omega_{Dj} \approx 0$. Therefore, for unstable modes with no trapped particles it is needed that $k_\parallel \neq 0$. In toroidal plasma for the majority of the trapped ions one has $\omega_{Di} > 0$, for the majority of trapped electrons, $\omega_{De} < 0$. Because of the differences in the velocities, the effect of the trapped electrons is more significant. According to (1), the trapped electrons can cause the instability when $k_\parallel \approx 0$. As calculated in [15], the growth rate γ of such modes is sharply reduced when the aspect ratio $R/a > 5$ (R is major radius, a is plasma minor radius). As the average aspect ratio of the L-2M stellarator is $R/a \sim 5$, the effect of the trapped particles can be neglected in the first approximation. Note that in the stellarator, the particles can be locally trapped due to inhomogeneity of magnetic configuration. Thus, we restrict our analysis to the most significant modes from the class of gradient drift instabilities: ITG and ETG. Given that the ratio of plasma pressure to magnetic pressure $\beta << 0.1$ in the L-2M, we will use the electrostatic approximation. In this case, the dispersion equation that combines the ITG and ETG modes has the form [15]:

$$\begin{aligned}
1+\left[1-\frac{\omega_{*e}}{\omega}\left(1-\frac{3}{2}\eta_e\right)\right]\xi_e Z(\xi_e)\Gamma_0(b_e) - \\
\frac{\omega_{*e}}{\omega}\eta_e\xi_e[\xi_e+\xi_e^2 Z(\xi_e)]\Gamma_0(b_e) - \frac{\omega_{*e}}{\omega}\eta_e\xi_e Z(\xi_e)\Gamma_0(b_e) - \\
\frac{\omega_{*e}}{\omega}\eta_e\xi_e Z(\xi_e)b_e[\Gamma_1(b_e)-\Gamma_0(b_e)] = \\
-\frac{T_e}{T_i}\left\{1+\left[1+\frac{\omega_{*i}}{\omega}\left(1-\frac{3}{2}\eta_i\right)\right]\xi_i Z(\xi_i)\Gamma_0(b_i) + \right. \\
\frac{\omega_{*i}}{\omega}\eta_i\xi_i[\xi_i+\xi_i^2 Z(\xi_i)]\Gamma_0(b_i) + \frac{\omega_{*i}}{\omega}\eta_i\xi_i Z(\xi_i)\Gamma_0(b_i) + \\
\left. \frac{\omega_{*i}}{\omega}\eta_i\xi_i Z(\xi_i)b_i\left[\Gamma_1(b_i)-\Gamma_0(b_i)\right]\right\}
\end{aligned} \tag{2}$$

Here $\omega_{*i} = k_\perp \dfrac{k_B T_i}{q_i B L_n}$, $\omega_{*e} = k_\perp \dfrac{k_B T_e}{e B L_n}$, k_B is the Boltzmann constant, T_i and T_e are ion and electron temperatures, q_i is the charge of the ions, e is the electron charge, B is the magnetic induction, L_n is the spatial scale of the density gradient, η_i and η_e are the relative gradients of ion and electron temperatures, $\Gamma_n(b) = I_n(b)\exp(-b)$, $I_n(b)$ is the modified Bessel function, $b_i = k_\perp^2 \rho_{Ti}^2$, $b_e = k_\perp^2 \rho_{Te}^2$, $Z(\xi) = \frac{1}{\sqrt{\pi}} \int_{-\infty}^{\infty} \frac{e^{-u^2} du}{u - \xi}$ is plasma dispersion function of argument $\xi_i = \frac{\omega}{k_{||}\sqrt{2k_B T_i / m_i}}$ (for ions) or $\xi_e = \frac{\omega}{k_{||}\sqrt{2k_B T_e / m_e}}$ (for electrons), m_i and m_e are ion and electron masses.

The analysis was performed for different regimes of plasma confinement in the L-2M with different levels of the EC heating power (200 and 400 kW). Figure 1 shows the model profiles of temperature, density and pressure for different modes used in the analysis taking into account the density pump-out effect. In figure. 1, r is the radial coordinate. The measured temperature profiles are well described by the quadratic parabola with a pedestal at the plasma boundary. The thickness of the boundary layer where the density drops to zero is approximately equal to 1 cm (the boundary layer where the density drops to zero is approximately 1 cm thick). The temperature of ions will be considered as a constant over the cross section, except for the region near the boundary. The regimes 1–3 correspond to different time instants of quasi-steady-state phase of the discharge at a heating power of 400 kW without the limiter. In the regime 4 (with the limiter), the heating power is 200 kW, and the density profile is well approximated by a sixth-order parabola.

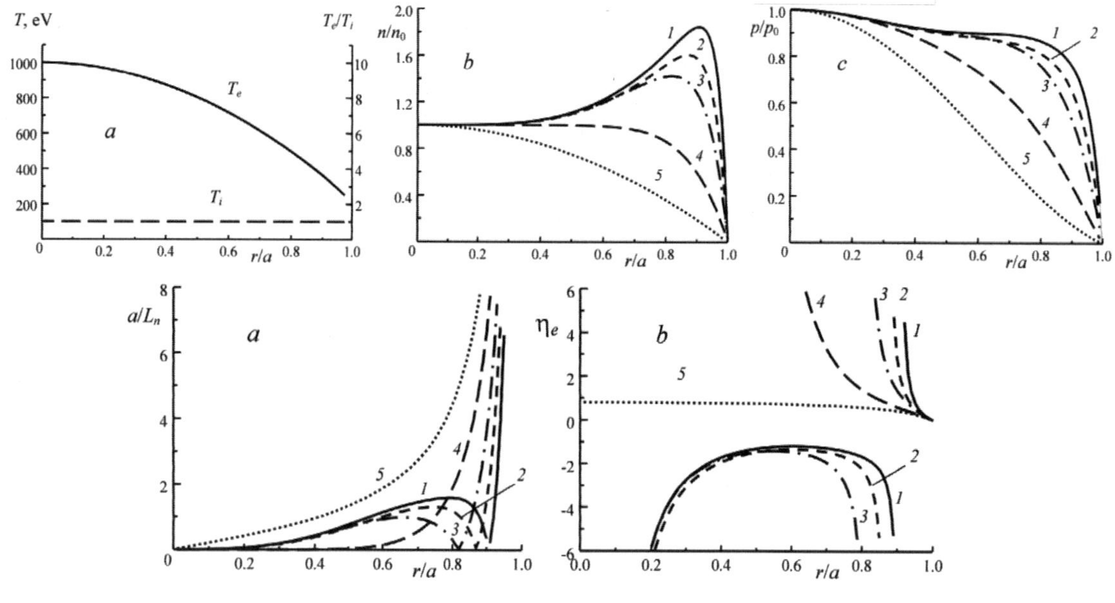

Figure 1. Top: profiles of temperature (*a*), density (*b*) and pressure (*c*) 1, 2, 3 – at different time instants at quasi-stationary ("flat-top") phase of discharge in the density pump-out operation mode at high heating power; 4 – 100 kW mode with low heating power of 100 kW (sixth order parabola); 5 – second-order parabola (given for comparison). Bottom: relative density gradient (*a*) and the ratio of gradients of electron temperature and density (*b*) for the regimes 1–5.

4. Evolution of the Fourier spectra of fluctuations during the plasma discharge

Figure 2 presents the evolution of the Fourier spectra of the plasma density fluctuations in the steady-state phase of the discharge. The dark line on the background of a noisy complex spectrum in each time frame shows the average (stable) spectra. At the quasi-stationary phase of the discharge within 54 to 62 ms interval at a constant average plasma density of $(1.8–1.9)\cdot 10^{13}$ cm^{-3}. The shifted broadband noisy spectra correspond to the steady-state of the discharge. The spectra half-width and intensity, recovered from the averaged values, vary but only slightly (frequency band variation is 20%, the spectral intensity variation is 30%).

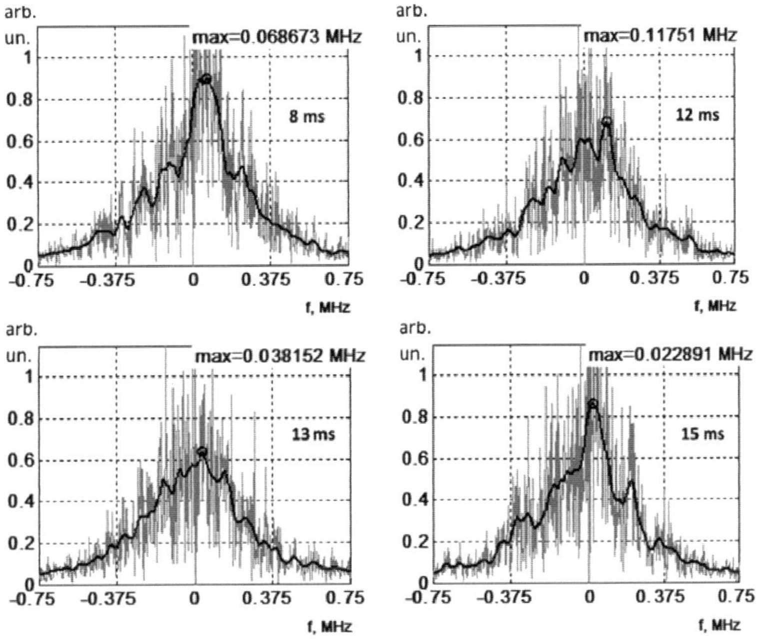

Figure 2. The evolution of the Fourier spectrum (gray) and the stable Fourier spectrum (black line) of plasma density fluctuations during the discharge in the L-2M stellarator. The time frame for spectrum estimations is 1 ms. The power of the EC heating is 400 kW. Shot number is # 19046 (43491).

Thus, an optimized Doppler reflectometer enabled us to obtain a repetitive complex-valued Fourier spectrum in the steady-state stage of the plasma discharge in the ECR heating mode of L-2M stellarator. The last two spectra in figure 2 (13 ms and 15 ms) correspond to the evolution of fluctuations in the end of discharge with the plasma column rotation velocity of about 150 m/s (the shift is evaluated by the local maximum of the spectrum). Note that in the mode with the high power heating during the formation of the density pump-out in the plasma core, the Doppler shift of the spectrum is reduced compared to the modes with a lower power and a parabolic profile of plasma density [3].

In the course of the research it was found that reducing the error when measuring the spectra with the optimized DR diagnostics has not led to improvements in the poloidal rotation velocity estimation, compared to the non-optimized reflectometer measurement. Figure 2 shows the local maximum value and the corresponding Doppler frequency shift of Fourier spectra at each time frame. Assuming that the poloidal rotation velocity of fluctuations exceeds the phase velocity, this frequency shift may be associated with the velocity of plasma poloidal rotation. Assuming that the radial electric field, which determines the poloidal rotation velocity of the plasma at the steady-state stage, is constant, the Doppler frequency shift is expected to remain the same for a given accuracy of measurement of the spectra. However, the Doppler shift in the 12 ms spectrum is 0.117 MHz, while in the 15 ms spectrum is 0.022 MHz. The Doppler shift, estimated from the spectra's maximum, varies almost by an order of magnitude at the steady-state stage of the discharge. This may indicate that the Doppler frequency shift is determined not only by the poloidal rotation velocity, but also by its mixture with the phase velocities of the turbulent fluctuations (provided the velocities are of the same order of magnitude). In this case, to determine the poloidal velocity it is necessary to highlight the harmonics associated with rotation of the plasma on the background of harmonics determined by the phase velocities of the turbulence in the complex-valued spectrum of the Doppler reflectometer signal.

Figure 3 shows the decomposition of the Fourier spectra of fluctuations of the measured DR signal on the components using the method presented in [16, 17]. The decomposition of the two spectra in the stationary stage of the discharge in 52–54 ms and 58–60 ms time windows is given. Initially six processes are set. In the considered spectral range we can distinguish the three processes, exceeding the experimental noise, which has characteristic values of the Doppler frequency shift. Three harmonics are clearly distinct. Two additional modes, besides harmonics related to the plasma rotation, correspond to oscillations propagating in the opposite directions: phase velocity of the first oscillation is directed towards the electron diamagnetic drift, the phase velocity of the second oscillation is directed towards the ion diamagnetic drift.

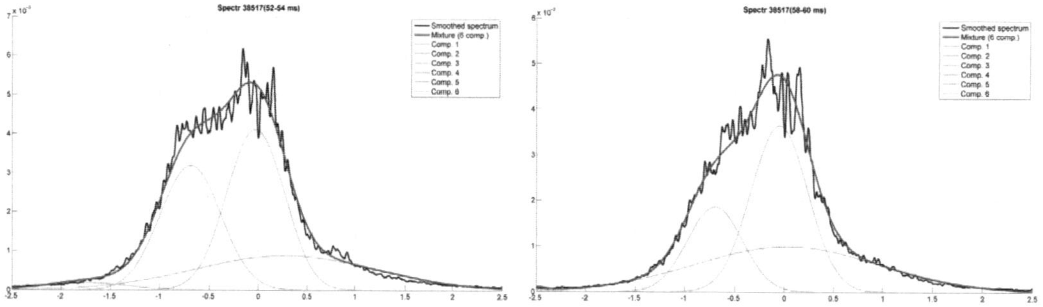

Figure 3. Decomposition of the Fourier spectra of fluctuations of the measured DR signal. The power of the EC heating is 400 kW.

Figure 4 shows the relative scale of the density gradient a/L_n and the relative gradient of the electron temperature η_e, that is needed to calculate the development of the ETG and ITG modes in the edge plasma of the L-2M. The ion temperature gradient was assumed to be zero ($\eta_i = 0$) to calculate the ETG mode. The results of the calculations are shown in figure 4. The values of $\omega_0 = k_B T_i/(eBa\rho_{Ti}) = 8.5\cdot10^5$ rad/s ($f_0 = 135$ kHz), and $\rho_{Ti} = 0.75$ mm were taken as the reference values. Higher values of the increments correspond to peripheral regions of the plasma column. For the calculation of the ITG-mode the presence of a gradient of ion temperature at the plasma periphery was assumed. We used the relative gradient with $\eta_i = 3$.

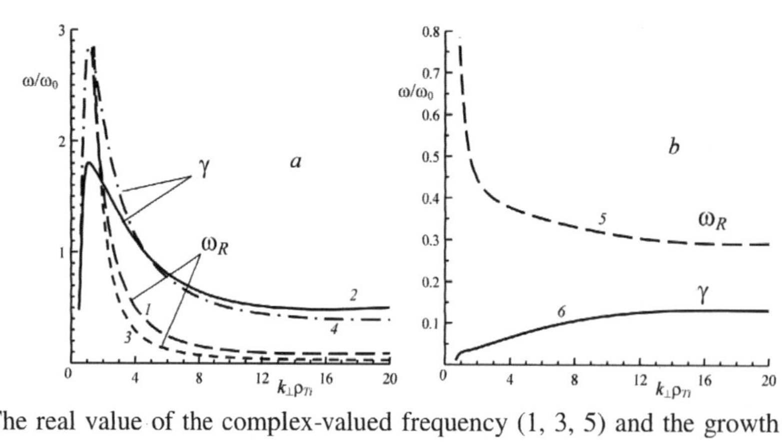

Figure 4. The real value of the complex-valued frequency (1, 3, 5) and the growth rate (2, 4, 6) of ETG instability versus transverse wave number at $r/a = 0.95$ (*a*) and $r/a = 0.6$ (*b*) for $\eta_i = 0$, $k_{\|}a = 0.05$: 1, 2 – pump-out profile; 3, 4 – sixth-order parabola. The scales are as follows: $\omega_0 = k_B T_i/(eBa\rho_{Ti}) = 8.5\cdot10^5$ rad/s ($f_0 = 135$ kHz), $\rho_{Ti} = 0.7$ mm.

The results of calculations are presented in [17]. Under the above conditions the typical increments of the ion and electron modes are close within the order of magnitude. Thus, in the L-2M stellarator, the ITG and ETG modes can be observed simultaneously. Let's estimate typical values of mode frequency for the detecting wave number $k_{\perp} \approx 2$ cm^{-1} in the coordinate system rotating in poloidal direction along with the plasma. The frequency shift relative to the poloidal rotation frequency is of the order of 1 MHz for the ETG mode, and 100 kHz for the ITG mode. The estimated values are in a qualitative agreement with the data presented in figure 2.

5. Conclusion

The analysis of the fluctuations spectra and instabilities was performed for various regimes of plasma confinement in the L-2M stellarator at electron-cyclotron heating power of 200 and 400 kW. It is theoretically shown that the development of the ETG and ITG instabilities in the edge of the stellarator plasma for the regimes with a high heating power and the formation of a dip on the density profile in the plasma core (the density pump-out effect) are possible. In all the spectra analysed, one can distinguish the components associated with the poloidal plasma rotation determined by the radial electric field, and the modes of structural turbulence of two types (determined by the ITG and ETG instabilities). It was found out that to reduce the error of the optimized Doppler reflectometry diagnostics of the plasma rotation poloidal velocity it is necessary to identify the harmonic of the complex-valued spectrum, related to the rotation of the plasma, on the background of harmonics determined by the turbulence's phase velocities. In this case, the local measurements of spectra using the Doppler reflectometry diagnostics enables one not only to measure the plasma rotational velocity, but also to describe the development of low-frequency plasma instabilities.

6. Acknowledgment

Work was partially supported by the Ministry of Education and Science of the Russian Federation, the contract No. 13.2573.2014/K

References

[1] Korolev V Yu and Skvortsova N N 2006 *Stochastic Models of Structural Plasma Turbulence* (Utrecht: VSP)
[2] Batanov G M et al 2013 *JETP Letters* **78** 502
[3] Akulina D K et al 2008 *Plasma Physics Reports* **34** 1059
[4] Conway G D 2008 *Plasma Phys. Control. Fusion* **50** 124026
[5] Chirkov A Yu and Khvesyuk V I 2011 *Plasma Physics Reports* **37** 437
[6] Kharchev N K et al 2011 *Plasma and Fusion Research* **6** 2402142
[7] Letunov A A et al 2014 *Proc. of XLI Int. Conf. of Plasma Phys. and Controlled Fusion* (Zvenigorod) 73
[8] Pshenichnikov A A et al 2005 *Plasma Physics Reports* **31** 554
[9] Chernov N A et al 2013 *Proc. of XXXIX Int. Conf. of Plasma Phys. and Controlled Fusion* (Zvenigorod) 68
[10] Mikhailovskii A B 1977 *Theory of Plasma Instabilities* **2**
[11] Rhodes T L et al 2007 *Plasma Phys. Control. Fusion* **49** B183
[12] Chirkov A Yu and Khvesyuk V I 2010 *Phys. Plasmas* **17** 01210
[13] Jenko F et al 2000 *Phys. Plasmas* **7** 1904
[14] Chirkov A Yu 2011 *J. Fusion Energy* **33** 139
[15] Lee Y C et al 1987 *Phys. Fluids* **30** 1331
[16] Gorshenin A K et al 2011 *Mathematical modelling* **23** 83
[17] Malakhov D et al 2014 *Proc. XXXII Int. Seminar on Stability Problems for Stochastic Models* (Trondheim) 68

Journal of Physics: Conference Series **666** (2016) 012008 doi:10.1088/1742-6596/666/1/012008

Experimental study of Z-pinch radiation spectra at the Angara-5-1 facility

V V Alexandrov, G S Volkov, E V Grabovsky, A N Gritsuk, I G Malyutin[1], K N Mitrofanov and G M Oleinik
State Research Center of Russian Federation "Troitsk Institute of Innovative and Thermonuclear Research", Pushkovyh street 12, Troitsk, 142190, Russian Federation

E-mail: malytin92@gmail.com

Abstract. The results of measuring the energy, power and spectrum of the soft X-ray radiation (SXR) pulse at the Angara-5-1 facility in the photon energy range of 0.02 – 2 kV are presented. A comparison of the SXR emitted in axial and radial directions is presented.

This paper presents the results of measuring the energy, power and spectrum of the soft X-ray radiation (SXR) pulse in the photon energy range of 0.02 – 2 kV, observed in the powerful Z-pinch in axial and radial directions. The data was obtained by analyzing the implosions of cylinders (1.2 cm in diameter and 1.6 cm high) composed of 6 μm-diameter tungsten wires with linear mass of 220 μg/cm at the Angara-5-1 facility in the range of the discharge current of 2.2 – 3 MA.

Two sets of vacuum X-ray diode (VXRD) sensors were used to measure temporal profile of the SXR radiation power as well as to reconstruct spectral features of powerful Z-pinch radiation in axial and radial direction. The radial set consisted of 4 VXRDs placed behind various filters while the axial one consisted of 3 VXRDs behind the same filters. Spectral range of each VXRD sensor was determined by quantum efficiency curve of the photocathode used and by spectral dependence of transmission factor of the set of filters made from different materials [1].

The spectral characteristics of the SXR radiation source during implosion of the Z-pinch based on the tungsten (wolfram) cylindrical wire liners (W-CWL) also was studied using the diffraction grazing incidence spectrometer (DGIS) with the off-Rowland circle registration scheme of spectrum registration. Figure 1 shows the optical scheme of the DGIS:

[1] To whom any correspondence should be addressed.

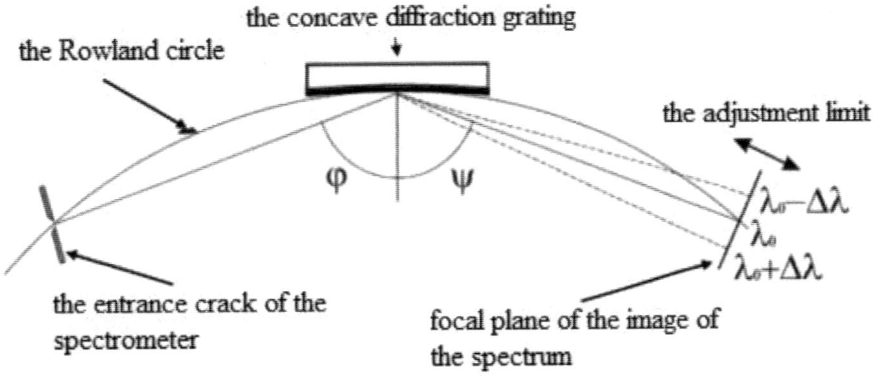

Figure 1. Optical scheme of the diffraction grazing incidence spectrometer (DGIS) with the off-Rowland circle scheme of spectrum registration (φ- angle of incidence of the studied X-ray flux on the concave diffraction grating; ψ- the diffraction angle of the x-rays of a wavelength λ_0; $\Delta\lambda$ - the wavelength range of the spectrum registration.

The high power and spectral brightness of the SXR radiation source makes it possible to register the spectra with a spatial resolution on the UV-4 X-ray film using the DGIS located inside the vacuum chamber at a distance of 1.5 m from the source. To reduce the detector illumination background signal we provided a shielding of the spectrometer and detector elements using the collimators and the lead shields of 1 - 2.5 cm thickness.

Experimental data on the temporal profiles of the SXR radiation pulse power, during implosion of the W-CWL, in the following ranges of photon energy, 20 - 70 eV, 70 - 300 eV, 300 - 1500 eV, as well as the spectrum-integrated power, were recorded by the sets of the VXRD sets) in the radial and axial directions of observation. These data allow to solve the inverse problem of recovering the spectral power of the SXR radiation at any time instant [2].

Figure 2 shows the dependence of an average spectral energy density (SED) of the SXR radiation of the W-CWL Z-pinch on the energy of photon emitted in the radial and axial directions.

These dependencies show that the radiation of the Z-pinch plasma for the given load (W-CWL: 12 mm diameter multiwire array of the forty W 6 μm diameter wires) in the radial direction is harder than that in the axial one. I.e., in the radial direction the main part of energy is emitted in the range of 70 – 300 eV, whereas in the axial direction the energy is distributed more uniformly, in the ranges of 20 - 70 eV and 70 – 300 eV.

Thus, the intensity distribution of the cylindrical Z-pinch source of the SXR-radiation is isotropic in height and azimuthal angle, and spectrum of the radiation in the axial direction corresponds to a softer spectrum compared to that in the radial direction.

Journal of Physics: Conference Series **666** (2016) 012008 doi:10.1088/1742-6596/666/1/012008

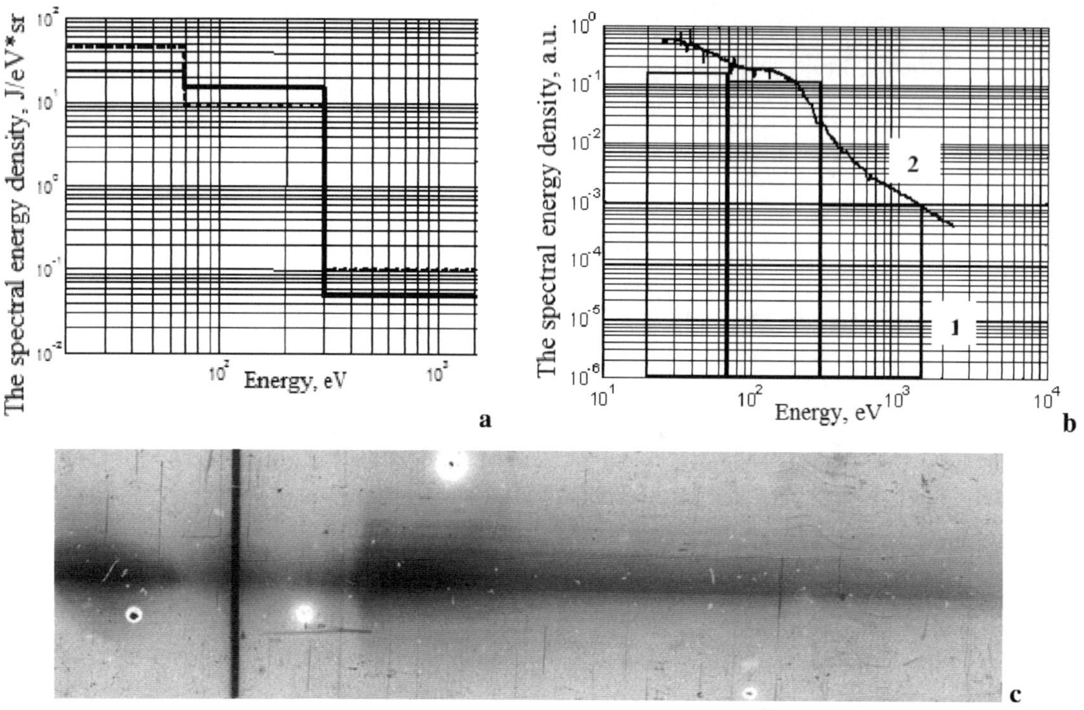

Figure 2. (a) Dependence of the average SED of the SXR radiation, emitted by the W-CWL Z-pinch in the radial and axial directions, on the photon energy (solid line - SED_{radial}^{avg}; dashed line - SED_{axial}^{avg}); (b) Comparison of the SED of the SXR radiation from the central region of the Z-pinch in the radial direction, evaluated with different methods (1 - SED data from VXRD, 2 - SED data from DGIS); (c) An image of the spectrum of the SXR radiation of tungsten cylindrical pinch with a spatial resolution along its radius for a shot 5274.

This work was supported in part by the State Atomic Energy Corporation ROSATOM under Contract H.4X.44.90.13.1108 and in part by the Russian Foundation for the Basic Research under Project 12-02-00900, 12-02-00369, 13-02-00013, 13-02-00482, and 14-02-00438.

References

[1] http://henke.lbl.gov/optical_constants/index.html

[2] Branitskii A V and Oleinik G M 2000 Reconstruction of the Parameters of a Soft X-radiation Spectrum from Signals of Vacuum X-ray Diodes *Instruments and Experimental Techniques* **43** 486

[3] Mitrofanov K N, Grabovski E V, Gritsuk A N, Laukhin Ya N, Aleksandrov V V, Oleinik G M, Medovshchikov S F and Shevel'ko A P 2013 Specific features of the structure of the Z-pinch emitting region formed during the implosion of a foam-wire load at the ANGARA-5-1 facility *Plasma Physics Reports* **39** 62

Wavelet analysis of the parameters of edge plasma fluctuations in the L-2M stellarator

S A Maslov[1,4], **D G Vasilkov**[2], **Yu V Kholnov**[2] **and N N Skvortsova**[2,3]
[1] Joint Institute for High Temperatures of the Russian Academy of Sciences (JIHT RAS), Izhorskaya st. 13 Bd.2,Moscow, 125412, Russian Federation
[2] A.M. Prokhorov General Physics Institute of the Russian Academy of Sciences, Vavilova Street 38, Moscow, 119991 Russian Federation
[3] National Research Nuclear University MEPhI, Kashirskoe shosse 31, Moscow, 115409, Russian Federation

E-mail: sergm90@mail.ru

Abstract. Wavelet analysis results are presented for evolution of the spectral fluctuation characteristics of the edge plasma density and potential in the L-2M stellarator for fast and slow transport transitions. The fast transition comes out as a sharp increase of the energy and electron density within ~0.1ms and the slow one as a weak parameter change during a 0.5 to 1 ms time interval. It is shown that the use of the "Mexican hat" and Morlet wavelets allows one to detect the moment of the fast transition, whereas applying the Haar wavelet adds to this also an estimate of its duration, conforming to the analytical calculations, and reveals the temporal structure of the slow transition.

1. Introduction

Transport transitions (TT) in the L-2M at the stationary discharge stage were first discovered in experiments with the maximum heating power of 250 kW [1]. They appeared as short-time processes being accompanied with a sudden increase of the energy and electron density by 15% (or by 50% with the 500 kW heating power) and a reduction of the plasma potential and density fluctuations at the plasma edge.

This paper presents the wavelet analysis of the plasma edge fluctuations for the fast and slow TTs. The fast TTs are the ones with the duration significantly shorter than the energy confinement time of the plasma. If the transition duration and the energy confinement time are comparable, then such TTs are called slow. Wavelet analysis allows detecting the temporal changes in the process parameters [3] and that is why it is suitable for the time sample research in different plasma diagnostics: dispersion, interferometry, Doppler reflectometry etc. It is shown that the Morlet wavelet and the "Mexican hat" [4] allow to detect the moment of the fast TT, whereas applying the Haar wavelet [4] adds to this also evaluation of its duration in a good accordance with the theory [2] and reflects the slow TT characteristics in more detail than with the two other functions.

[4] To whom any correspondence should be addressed

Journal of Physics: Conference Series **666** (2016) 012009 doi:10.1088/1742-6596/666/1/012009

2. Description of the experiment and signal processing

Plasma generation and heating in the L-2M stellarator [1, 2] were realized with the ECR method using the gyrotron complex MIG-3 [5]. In the experiments, the maximum power was 500 kW. In order to study the plasma edge characteristics (at the immersion depth less than 1 cm), the L-2M device employs the movable Langmuir probe system configured to measure the floating potential V_f and the saturation current I_{sat}. Experimental data recording to the processing device occurs at the 100 kHz sampling frequency for the signals with fast TT and 1 MHz for the slow TT. The MATLAB software is used to construct the spectra.

Wavelet analysis is carried out for the V_f and I_{sat} experimental data obtained from the Langmuir probes. The signals are analyzed in the time interval 52 to 60 ms using three basic wavelets: the "Mexican hat" and the Morlet and Haar functions. The first two wavelets have a good spatial and temporal localization, so they are widely applied to the analysis of plasma signals in stellarators and tokamaks [6, 7].

The Haar wavelet is a discontinuous function with a much worse localization than the other ones, so it is used more rarely for the plasma signal analysis, especially for processes with a smooth temporal data evolution. One of the few applications of this wavelet is the analysis of the MHD turbulence in the solar wind [8]. However, as will be shown in Section 3, applying the Haar wavelet can yield more significant results than using the Morlet or "Mexican hat" functions.

In order to isolate the fluctuation components of the signals, the low frequency filtering procedure is performed: at every time moment, the values of the floating potential and the Langmuir probe current averaged over next 25 time points recorded are subtracted from the signals. Then the scalograms are built for the floating potential and the saturation current. For this purpose the signal wavelet transformation is calculated:

$$W_f(a,b) = \frac{1}{|a|^{1/2}} \int_{-\infty}^{+\infty} f(t)\psi\left(\frac{t-b}{a}\right)\mathrm{d}t \quad (1)$$

where $f(t)$ is the signal and $\psi(t)$ is the basic wavelet [4] with the width a and the localization center b. The scalogram is built in the parameter plane (a,b) and the value of $C(a,b) = |W_f(a,b)|^2$ is indicated with gray hue in the scalogram (the lighter the shade, the higher the energy).

The wavelet spectrum [9] reflects the variation of the energy of the signal fluctuations in time, i.e. it is a function not of a and b, but of the frequency ω and time t. The value of this function evolves in time in accordance with the change of the signal characteristics. The correlation between V_f and I_{sat} is also studied in this paper. This characteristic is evaluated using the mutual wavelet spectrum. The mutual, or the cross-spectrum [3, 10, 11] of the signals $f(t)$ and $g(t)$ on a finite time interval T is the function

$$B_{fg}^{W}(a,\Delta t) = \int_T W_f^*(a,t)W_g(a,t+\Delta t)\mathrm{d}t \quad (2)$$

In the graphs, the spectrum and cross-spectrum values are indicated with the shades of gray.

3. Experimental results

In the L-2 M experiments, the fast TTs were registered in several pulses. The fast transitions bore a great resemblance to the processes observed in [2, 12, 13] including the classic L-H transitions [14]. In the fast TT, the plasma energy and density grow quickly and then saturate (figure 1), and the density parameter remains constant even after turning off the ECR-heating; the H_α radiation intensity decreases and that of B_2 increases [1]. For the signal #16081, the transition is detected approximately at 57 ms and takes about 0.1 ms [2]. In the figures, the fast TTs moments are marked with the dashed line.

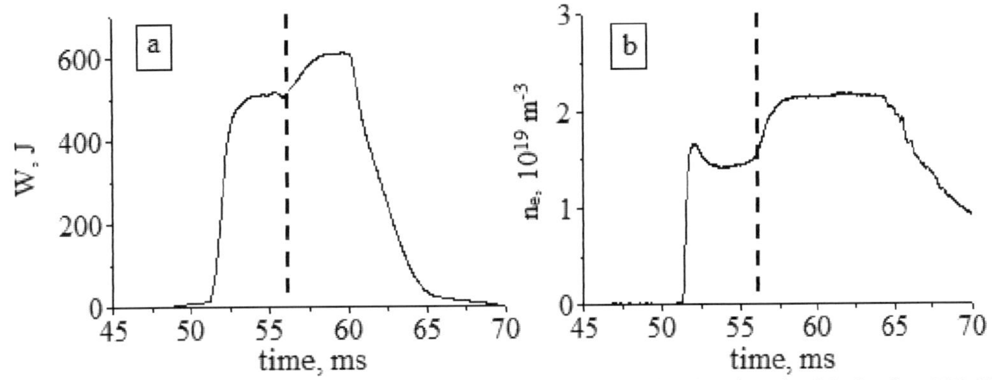

Figure 1. Energy (a) and electron density (b) of the plasma for the signal with the fast TT. Dashed line marks the transition moment.

In the slow TT, one can observe a different kind of evolution of the energy and electron density. One sees from figure 2 that before the transition, the plasma density can either decrease (signal #19124) or increase (signal #19166), but after the TT the density slightly increases. For the discharges #19124 and #19166, the TTs happen at time intervals 56.5 to 57 ms and 55.5 to 56 ms, correspondingly. The start and end moments of the slow TT are marked with the dashed lines.

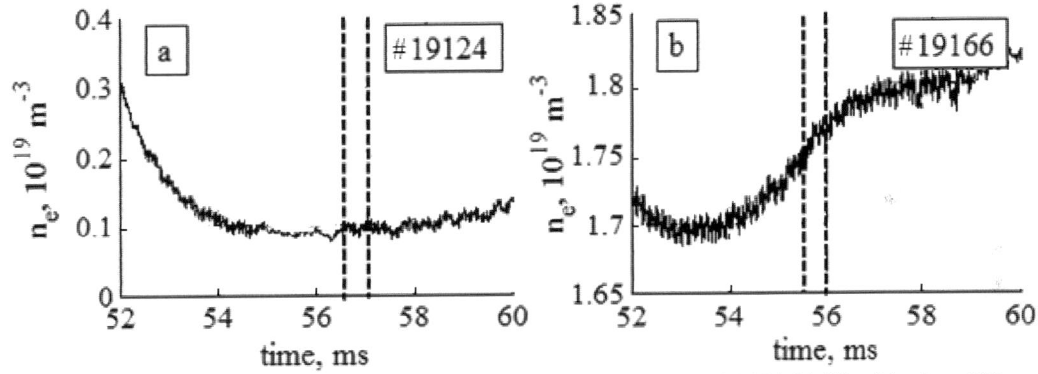

Figure 2. Plasma electron density in discharges #19124 (a) and #19166 (b) with slow TT.

Langmuir probes also detect the change of the signal parameters during the TT. In the case of the fast TT, the V_f and I_{sat} fluctuations decrease by an order of magnitude along with the increase of the plasma density. During and after the slow TT one can observe a slight growth of the amplitude of the potential fluctuations. During the TT the local maxima of the V_f fluctuations are recorded at 56.62 ms, 56.81 ms and 56.91 ms and the minima at 56.70 ms and 56.88 ms. Figure 3 shows the evolution of the floating potential fluctuations for the fast (discharge #16081) and slow (discharge #19124) transitions.

Journal of Physics: Conference Series **666** (2016) 012009 doi:10.1088/1742-6596/666/1/012009

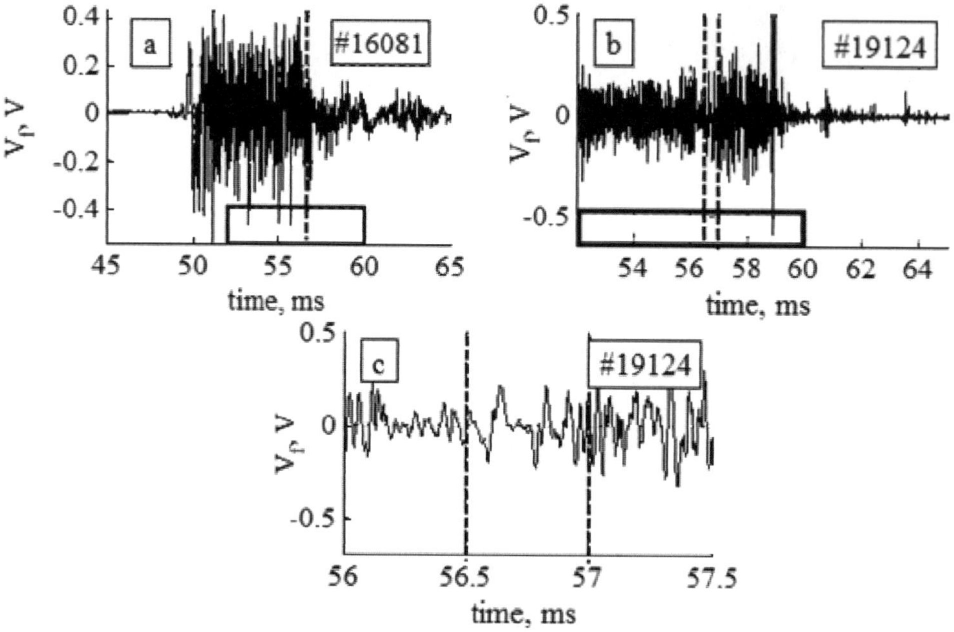

Figure 3. Edge plasma potential fluctuation for the signals #16081 (a), #19124 (b), #19124 in the interval 56 to 57.5 ms (c). The rectangle outlines the time interval analyzed.

Evolution of the plasma edge characteristics after the TT is presented as the wavelet scalograms. For the signal #16081 the scalograms built with the "Mexican hat" and the Morlet wavelet show weakening fluctuations of the floating potential at 57 ms (figure 4) and that of the current at 56.5 ms – approximately at the same time with the change of the plasma energy and electron density. The signal #19124 with a slow TT can also be characterized by the simultaneously changing electron density and floating potential at 56.5 to 57 ms. However, application of the continuous wavelets does not allow estimation of the fast TT duration, nor analyzing the temporal structure of the slow TT in detail.

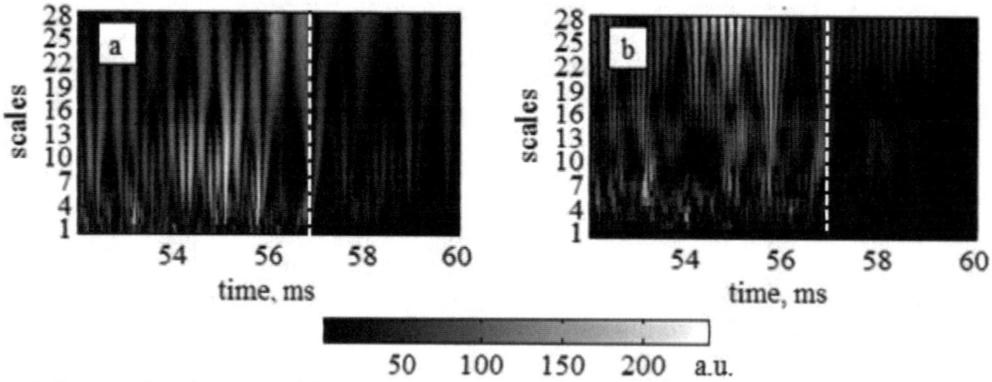

Figure 4. Potential scalograms of the signal #16081, built using the “Mexican hat” (a) and the Morlet function (b).

At the same time, the Haar wavelet (figure 5) allows to solve these problems. The resulting fast TT duration estimate of 0.1 ms shows good accordance with the theory. Scalograms built with the Haar wavelet reflect different behaviour of the potential and current of the signal #16081 after the transition. The potential fluctuations damp instantly at the TT moment and slightly increase at 58 ms, whereas the current fluctuations after the sharp reduction at 56.5 ms decrease gradually for 0.5 ms and then

Journal of Physics: Conference Series **666** (2016) 012009 doi:10.1088/1742-6596/666/1/012009

sustain the average value up to 60 ms. Besides that, the Haar function allows to record all the extrema of the fluctuation amplitude during the slow TT, which is not the case with the continuous wavelets.

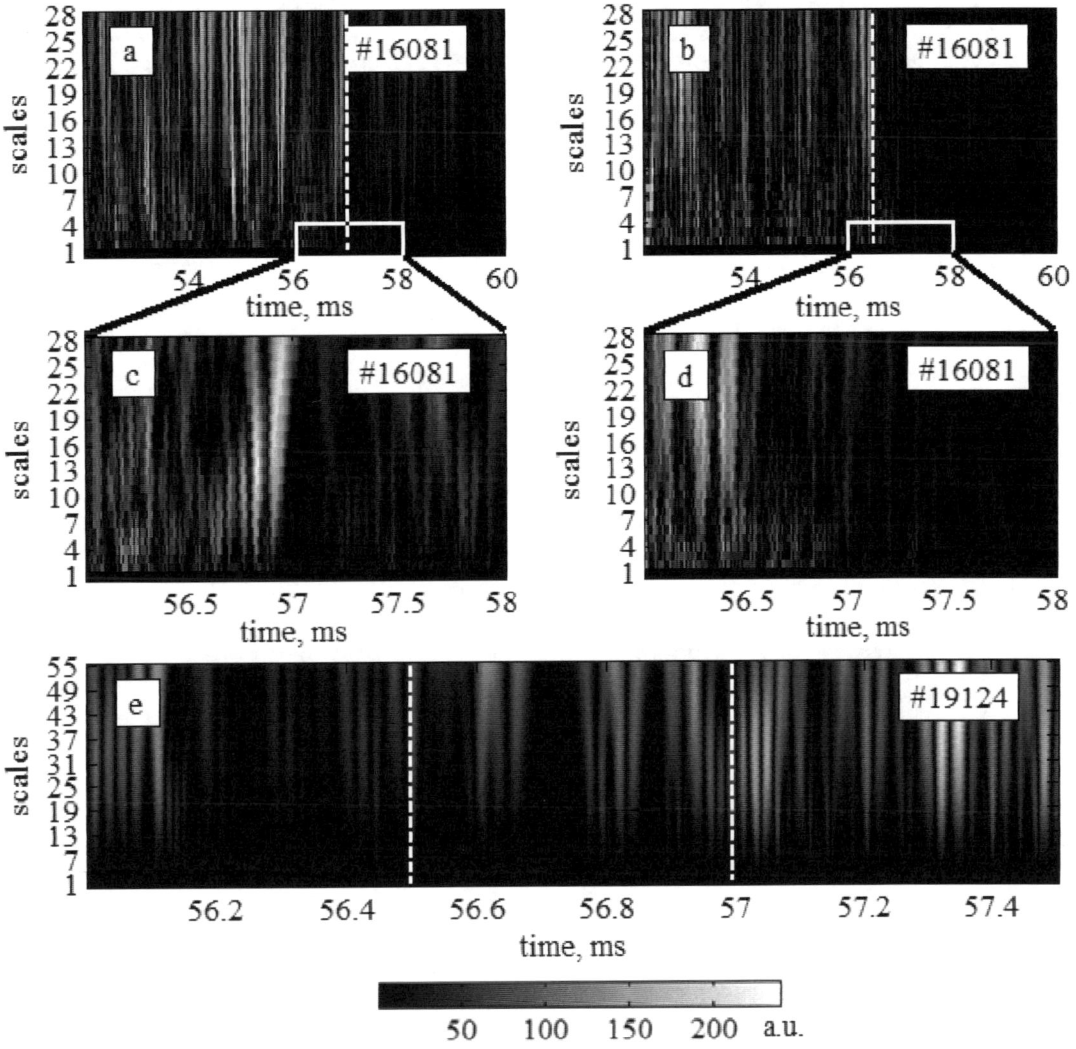

Figure 5. Signal scalograms built using the Haar wavelet: the potential for the signal #16081 (a, c), the current in the discharge #16081 (b, d), the potential in the discharge #19124 (e).

For the signal #16081 the wavelet spectra were also constructed. The analysis proved that before the fast TT, the fluctuations frequencies of V_f and I_{sat} in the range from 0 to 50 kHz are distributed approximately uniformly and after the transition one can see a difference. The energy of the potential fluctuations for the frequencies from 10 to 50kHz decreases sharply and for the lowest-frequency oscillations (less than 10 kHz) it changes insignificantly (figure 6a). The current spectrum for the signal #16081 remains approximately uniform, but its values are greatly reduced (figure 6b).

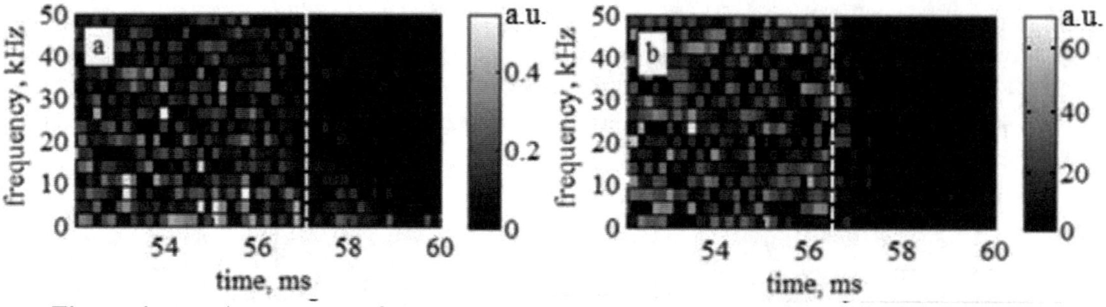

Figure 6. Wavelet-spectrum of potential (a) and current (b) of the signal #16081 with fast TT.

The mutual spectrum of the floating potential and current for the discharge #16081 was also built using the "Mexican hat", Morlet and Haar functions. Coherence maximums are recorded at 52.4 ms and 55.3 ms. The fast TT is accompanied by a sharp decrease of the coherence between V_f and I_{sat}. However, in the case of using the "Mexican hat" and Morlet functions, one can observe blurring of the mutual spectrum. Therefore no clear conclusion about the cross-correlation of the parameters at the TT moment can be made.

The graph constructed using the Haar wavelet (figure 7) shows that at the fast TT moment, the potential, current and their correlation decrease sharply, for much shorter time than the signal #16081 duration. In the time interval 56.4 to 57 ms the spectrum energy is slightly higher than after 57 ms, which is attributed to the time difference of the sharp current weakening and the potential fluctuations (56.5 and 57 ms respectively).

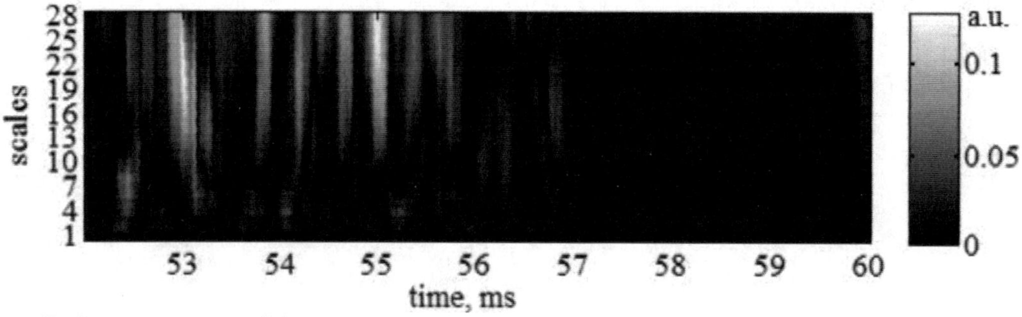

Figure 7. Cross-spectrum of the potential and current of the signal #16081 built with the Haar wavelet.

4. Conclusion

Spontaneous transport transitions are found in the L-2M stellarator experiments with the heating power P > 250 kW. The fast TTs are accompanied by a sharp rise in the energy and the plasma density during ~0.1 ms and a simultaneous drop of the fluctuations of the density and potential at the plasma edge. In the slow TTs, the plasma characteristics change simultaneously in 0.5 to 1 ms with a slight increase of the potential oscillations after the TT.

Wavelet scalograms reflect the variation of the amplitude of the noise component of the signal. It is the Haar wavelet only that provides also a reliable estimate of the fast transition duration and a detailed analysis of the temporal plasma edge parameters evolution in a slow TT.

The fast transition is accompanied by a sharp fall of cross-correlation of the potential and density parameters in the edge plasma. The smooth reduction of the coherence of the signal #16081 in the time interval 56.4 to 57 ms is caused by the difference of the transition moments for the potential and the current.

Acknowledgements

The study was financially supported by the RSF (Project №14-50-00124).

References

[1] Shchepetov S V *et al* 2008 *Plasma Phys. Control. Fusion* **50** 045001
[2] Kharchev N K *et al* 2011 *Plasma and Fusion Research* **6** 2402142
[3] Vasilkov D G *et al* 2014 *Proc. of XLI Int. Conf. of Plasma Physics and Controlled Fusion (Zvenigorod*) p 64
[4] Van Milligen B Ph *et al* 2005 *Phys. Plasmas* **2** 3017
[5] Daubechies I 2001 *Ten lectures on the wavelets* (Izhevsk, SRP: Regular and chaotic dynamics)
[6] Batanov G M *et al* 2003 *Plasma Phys. Reports* **29** 363
[7] Kharchev N K, Skvortsova N N and Sarksyan K A 2001 *J. Math. Sci.* **106** 2691
[8] Salem C *et al* 2009 *The Astrophysical Journal* **702** 537
[9] Wickerhauser M V 1991 *Proc.Ondelettes and Pacquettes d'Ondes (Rocquencourt)* p 31
[10] Torrence C and Compo G 1998 *Bull. Amer. Meteorol. Soc.* **79** 61
[11] Grinsted A, Moore J C and Jevrejeva S 2004 *Nonlinear Processes in Geophysics* **11** 561
[12] Letunov A A *et al* 2014 *Proc. of XLI Int. Conf. of Plasma Physics and Controlled Fusion (Zvenigorod*) p 73
[13] Shchepetov S V, Kholnov Yu V and Vasilkov D G 2010 *JETP Letters* **91** 170
[14] Wagner F *et al* 1982 *Phys. Rev. Lett.* **49** 1408

Journal of Physics: Conference Series **666** (2016) 012010 doi:10.1088/1742-6596/666/1/012010

Application of filter method for detection of secondary electron emission in the auto-oscillating mode of beam plasma discharge

A V Balovnev[1], I V Vizgalov and G H Salahutdinov
National Research Nuclear University MEPhI, Kashirskoe shosse 31, Moscow, 115409, Russian Federation

E-mail: alexisvb@mail.ru

Abstract. In this paper we studied the non-self mode of the auto-oscillation secondary-emission discharge (ASED) in a longitudinal magnetic field with autonomous electron gun to ignite the primary beam-plasma discharge (PPD).

Abnormally high secondary electron emission is the basis of a positive feedback mechanism leading to auto-oscillations at the resonant frequencies. For ASED [1], pulse amplitude of excited auto-oscillations can be much larger than the negative voltage applied to the collector plate, defining the features of the spectrum of high-energetic electron group. The experimental scheme is shown in figure1 [2].

Conventional contact probe and spectrometric research methods of diagnostics of distribution of high-energy group of electrons in the plasma column are carrying significant disturbance. Furthermore, they do not allow to separate the effects of primary and return flows including reflected and emitted by the surface of the electron collector [3].

We suggest the novel method of X-ray diagnostics of ASED, in which the spectrum of the high-energy (> 4 keV) electron group is reconstructed from X-ray radiation.

Our method uses a spectrometer based on thermoluminescent detectors (TLD) [4]. It provides 13 detection channels with TLD assemblies of lithium fluoride tablets arranged one after the other. Assemblies are closed with attenuation filters [5] of the aluminum foil. The type and thickness of the filters limits the sensitivity of the spectrometer (quantum energy > 4 keV). TLD indications were read with DVG-02TM device.

In the first experiments, the diaphragm has been deployed to the TLD thus sensors detected radiation from the primary beam after bombarding a target with primary electron flow (scheme in figure 1a). This PPR mode was obtained at an electron gun accelerating voltage of 5 kV (current 0.1A) without working gas inlet (chamber pressure of 10^{-6} Torr). Interaction of the primary beam with hydrogen plasma (hydrogen inlet, chamber pressure 10^{-4} Torr) leads to the collision less braking of the main part of electrons and acceleration of the rest electrons (about one percent) over the energy of the primary beam. These electrons provide the high energy "tail" in the distribution.

In subsequent experiments, measurements were made in the ASED mode. As shown in figure 1b, primary electron beam 1 passes through a hole in the diaphragm 4. The magnetic axis at the outlet of

[1] To whom any correspondence should be addressed

the trap has a slight deviation from the geometric axis. It can be configured by a special side coil.

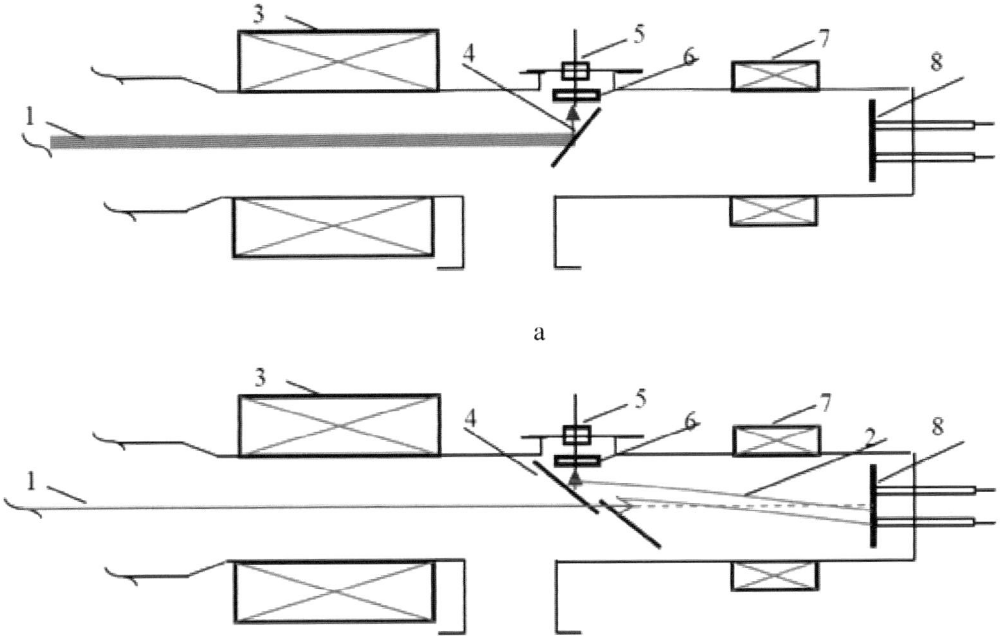

a

b

Figure 1 PPD measurements scheme (a), ASED measurements scheme (b). 1–primary beam; 2 – reverse flow of secondary and reflected electrons; 3,7–magnetic field coils; 4 –molybdenum diaphragm; 5 –locking device; 6–TLD cartridge ; 8 –collector plate.

When collector 8 is grounded, plasma column with the primary beam follows this rounding, since the polarization of the column under the influence of centrifugal drift is suppressed by a good electrical contact between the plasma column and the collector by a thermal electron component.

When collector potential is biased negatively with an external source, contact for electronic components is broken and flute instability is developed [3], resulting in the flattening of the plasma column. This leads to separation of the primary and secondary-emissive reverse flow - reverse electron flow is experiencing a drift in crossed magnetic and induced electric fields. It moves along the radius of curvature of the magnetic tube and falls on the solid area of the diaphragm that allows to register accompanying X-ray radiation.

As a result of the spatial separation, the X-rays of the secondary electron emission stream is detected by our sensor.

The accelerating voltage of the gun was 1.5 kV, gun emission current 0.1A, the pressure was the same as in the previous experiment - 10^{-4} Torr, the bias voltage of the collector plate - 600V, the amplitude of auto-oscillations obtained - 600V, frequency - 3 MHz. The energy of secondary-emissive electrons should not exceed 1200 eV, but the computing of the absorption curve indicates the presence of the abnormal, much harder X-ray component. Figure 2 summarizes the areas with the spectra obtained in the PPD and ASED modes.

Their comparison shows that in PPD mode a share of high energy accelerated electrons is very small, as indicated by a sharp decline of relative intensity of X-radiation when primary electron energy is exceeded (5 keV). In ASED mode X-ray radiation spectrum differs in principle - its spectrum reveals much larger relative number of multiply accelerated electrons.

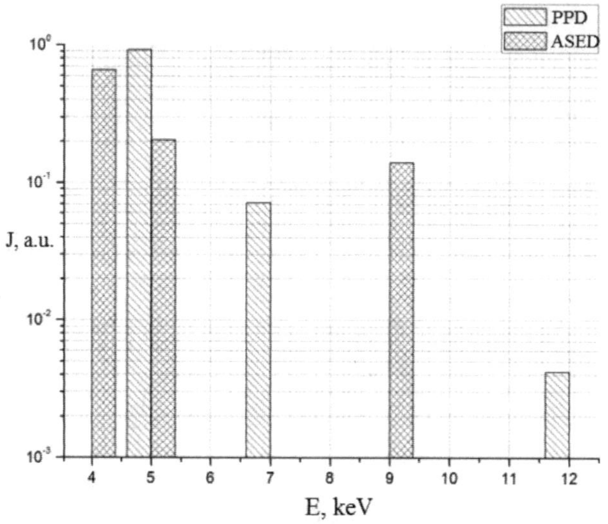

Figure 2. Spectra obtained in PPD and ASED modes.

Note that in the beam-plasma interaction a small proportion of primary electrons can be accelerated and confined between the plugs, gaining quite a lot of energy. In ASED mode experiments these electrons are captured either in a collector located in a deep cavity or in an opposite to the sensor side of the diaphragm, and generated radiation is not registered. Thus, in ASED mode with spatial separation of direct and reverse flows additional acceleration should occur only once in a single passage between the emitter and the target.

This may be due to a grouping effect. The experimental and calculated collector voltage waveforms show that ASED mode can be characterized by the growth of the time derivative on the front edge of the negative voltage pulse. As a result of the spatial and temporal compression of the electron cluster in ASED mode the beam-plasma interaction can be more efficient compared to the conventional PPD mode, and thus empower accelerated electron groups.

In addition, the anomalous acceleration of the secondary electrons can be associated with a very high-frequency modes of auto-oscillations. As a result, the actual acceleration voltage can grow much higher than the measured potential of the receiving plate relatively to the installation body.

This work was partly supported by a grant of the President of the Russian Federation for state support of the leading science schools of the Russian Federation (NS-5084.2014.2).

References

[1] Akel M, Vizgalov I V and Kumaev V A 2003 Observation of a plasma surface instability in a linear edge plasma simulator *30th EPS Conference on Controlled Fusion and Plasma Physics, St Petersburg, Russia, July 7-11,2003* **P-3** 180
[2] Lebedev P M *et al* 1976 *Plasma Physics Reports* **2** 407
[3] Lowson J *The Physics of Charged-particle Beams* (Oxford, 1980)
[4] Shani G 2000 *Radiation Dosimetry Instrumentation and Methods* CRC Press 222
[5] Kushin V V, Lyapidevsky V K and Perezhogin V B *Nuclear Physics Methods of Plasma Diagnostics* (Moscow, 1985) [in Russian]

Polarimetric investigations of nonthermal electrons in a high-current vacuum spark

O A Bashutin[1] and A S Savjolov
National Research Nuclear University MEPhI (MEPhI), Kashirskoe shosse 31, Moscow, 115409 Russian Federation

E-mail: boa308@mail.ru

Abstract. The investigation of nonthermal electron beams in a high-current vacuum spark by Bragg polarimetry of continuous radiation from discharge plasma is reported. Movement features of electron beams inside and outside of contraction region of the discharge plasma are revealed.

1. Introduction

Among X-ray sources of various types for technological and diagnostic applications, a special place belongs to pinched discharges. A characteristic feature of these is that the radiation spectrum contains photons with energies significantly exceeding the difference of potentials between electrodes. The origin of this radiation is related to high-energy electron beams formed during local pinching of the plasma column to micron dimensions (micropinch). Information concerning the directional and energy characteristics of such electron beams is important for understanding the mechanism of their formation and controlling the radiation parameters.

This report presents the results of our investigation of nonthermal electron beams formed in plasma of a high-current low-inductance vacuum spark, which was based on an analysis of the polarization of X-ray emission from the discharge plasma.

2. Experimental setup

The experiments were performed on a Zona-2 micropinch machine (National Research Nuclear University MEPhI, Moscow). A discharge was occurred out between a pointed anode and a plane cathode separated by a gap of 5 mm. Both electrodes were made of iron. The residual pressure in the vacuum chamber was p ~ 3×10–5 Torr. At the discharge voltage of 15 kV, the amplitude of the discharge current reached 150 kA. The period of discharge current oscillations was 8.5 μs. The X-ray emission spectrum was measured by two identical FSSR-1D focusing spectrographs with a spatial resolution in the direction perpendicular to the dispersion plane. The spectrographs were arranged so that the dispersion plane of one spectrograph was perpendicular, while that of the other spectrograph was parallel to the discharge axis. The spectra were recorded on an X ray sensitive photographic film. For the adopted adjustment, the system monitored a spectral interval of 1.83–1.88 Å.

Densitograms of the emission spectra integrated over the entire discharge gap are presented in the figure 1. Symbols w and y indicate positions of the resonance and intercombination spectral lines of

[1] To whom any correspondence should be addressed.

helium-like iron ions, respectively; symbol j indicates the satellite of the resonance line. As can be seen, the characteristic lines of multiply charged ions are observed on the background of a rather intense continuum. The intensity of the continuum with a predominantly perpendicular component $I_\perp$ (curve 1) is higher than that of the parallel component $I_\parallel$ (curve 2). The spatial region of continuous emission occurs between the point of micropinch localization and the anode.

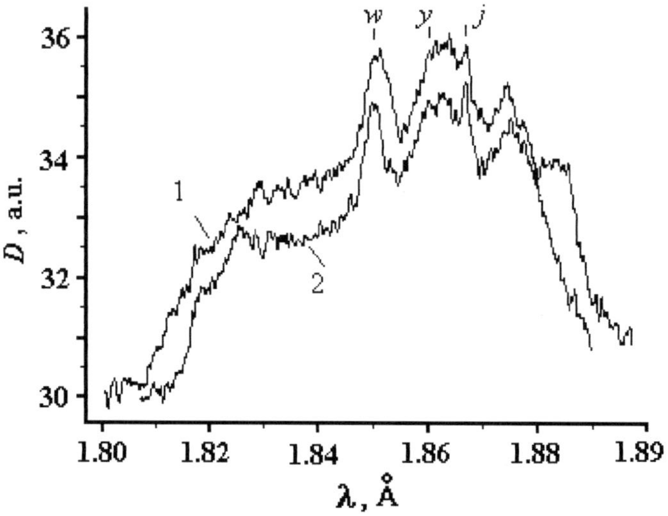

Figure 1. Integral densitograms of the emission spectra recorded by spectrographs with the dispersion planes oriented (*1*) parallel and (*2*) perpendicular to the discharge axis.

3. Discussion

The electron temperature in the micropinch reaches $T_e > 1$ keV. When electrons with these energies interact with atoms and ions in the surrounding plasma, the main contribution to continuous emission is due to bremsstrahlung radiation, which carries direct information on the directional motion of electrons. Model calculations [1] showed that the degree of polarization of the X-ray bremsstrahlung at a given electron beam energy E_0 can vary depending on the photon energy from $P_B(hv) = 1$ at low frequencies ($hv \to 0$) to $P_B(hv) = -1$ at high frequencies ($hv \to E_0$), in accordance with the following formula:

$$P_B(h\nu) = \frac{I_\perp(h\nu) - I_\parallel(h\nu)}{I_\perp(h\nu) + I_\parallel(h\nu)}$$

where $I_\perp(h\nu)$ and $I_\parallel(h\nu)$ are the intensities of emission components at frequency ν polarized perpendicular and parallel, respectively, to the discharge axis.

In the spectral interval studied ($h\nu \approx 6.6 - 6.7$ keV), the intensities of continuous emission components reconstructed from the results of measurements correspond to $P_B \approx 0.2$. That is, the given spectral interval corresponds to a low-frequency part of the bremsstrahlung radiation spectrum and, hence, is related to the axial flow of super thermal electrons with energies E_0 much greater than the energy of radiation photons detected by the spectrograph, i.e., with $E_0 >> 7$ keV. This estimation agrees with our previous data [2] about the presence of electrons with an equivalent temperature of $T_e > 20$ keV (for the Maxwell velocity distribution). The formation of a beam of super thermal electrons is probably related to the development of an anomalously high resistance of plasma in the region of pinching, accompanied by the appearance of a strong longitudinal electric field.

Analogous measurements were recently performed in [3]. Data obtained in the same spectral range were indicative of a negative polarization of bremsstrahlung within $-1 < P_B << 0$. This is evidence for the axial propagation of an electron beam with energy E_0 on the order of the energy of detected photons (i.e., E_0 ~ 8 keV), in agreement with data on the temperature of high-energy electrons determined for the same system by the "gray" filter technique. We believe that the low energy part of electrons in the axial beam was related to the used system of discharge initiation, which was located away from the discharge gap behind the anode. In this case, it is possible to form the current channels out of the compression region. This leads to decrease in the efficiency of pinching and, hence, in the magnitude of the generated axial electric field.

It has been proved in experiment that the polarization of bremsstrahlung emission from high-current low-inductance vacuum spark depends on the ratio of energies of the detected photons and beam electrons. This allows one to determine the energy of superthermal electrons from the degree of polarization of bremsstrahlung continuum in a given spectral range.

References

[1] Gluckstern R L and Hull M H 1953 *Phys. Rew.* **90** 1030
[2] Averin M S, Baykov A Y, Bashutin O A, Vovchenko E D, Dmitrusenko A S, Savjolov A S and San-Wei L 2006 *Instr. and exp. techniq.* **49** 265
[3] Dolgov A N, Klyachin N A and Prokhorovich D E 2014 *Plasma Phys. Rep.* **40** 733

Control of plasma parameters and spectral analysis sensitivity under bichromatic laser irradiation of materials in gases

A N Chumakov[1], N A Bosak and A V Panina
B.I. Stepanov Institute of Physics, National Academy of Sciences of Belarus, 68 Prospekt Nezavisimosti, Minsk BY-220072, Republic of Belarus

E-mail: chumakov@dragon.bas-net.by

Abstract. Double-pulse nanosecond bichromatic laser irradiation of carbon and brass targets is implemented. The dependence of temperature and density of electrons, as well as the specific recoil momentum of an outflowing plasma jet, on the ordering of laser pulses and the inter-pulse time delay is found. The results obtained can be used for the development of laser-plasma spacecraft microengines and the enhancement of laser spectral analysis' sensitivity.

1. Introduction

It is known that multi-pulse and double-pulse modes of laser exposure are more effective for local spectral analysis and for making the spacecraft laser-plasma engines as compared to single-pulse mode [1-7]. It was found that the double-pulse laser irradiation (LI) intensifies the rate of heating the targets' vapors within a certain range of time intervals between the pulses [2, 3, 7]. Furthermore, the hole drilling under laser pulses train action [4] and the laser-induced breakdown spectroscopy [5] were found out to be more efficient for these modes. Additional benefits in formation and heating of laser plasmas can be achieved by the double-pulse mode of the material irradiation at different wavelengths due to the dependence of the targets reflectance and radiation absorption of plasma on the wavelength of the LI.

2. Experimental setup

The double-pulse LI was produced by synchronizing two Nd:YAG lasers with wavelengths of 1064 and 532 nm (E = 100 mJ, τ = 20 ns) with adjustable time interval between laser pulses. The dependence of electron density and temperature of the plasma on the ordering of laser pulses and the inter-pulse time delay during laser exposure of various materials in air was studied. The most representative data were obtained using the brass and graphite targets.

In these experiments, the radiation emission spectra of a near-surface plasma and the pressure pulses on the irradiated targets were registered. The SDH-IV spectrometer with exposure time of 7 ms was used for spectral registration of laser plasma. The spectrometer was equipped with a linear CCD detector TCD 1304 AP (Toshiba) connected to a computer and a laser system synchronization unit. For registration of pressure pulses on the target and of the LI pulses, a piezoelectric detector synchronized with the Bordo 221 digital oscilloscope was used. Recoil momenta for different materials were determined by integrating the pressure signal over time.

[1] To whom any correspondence should be addressed.

3. Results and discussions

The structure of laser plasma plume and the spectra under the double-pulse LI of a brass target is presented in figure 1. Figure 1a shows the images of the brass plasma plume for different ordering of pulses and various time intervals between them. The photographs served as a basis for estimating the characteristic size of the luminous area of the laser plume, which varies from 8.5 to 11 mm depending on the ordering of laser pulses and the time intervals between them. In figure 1, the negative values of intervals between laser pulses correspond to the target's irradiation starting with the shortwave laser (λ=532 nm), and the positive ones correspond to a start with the longwave laser (λ = 1064 nm).

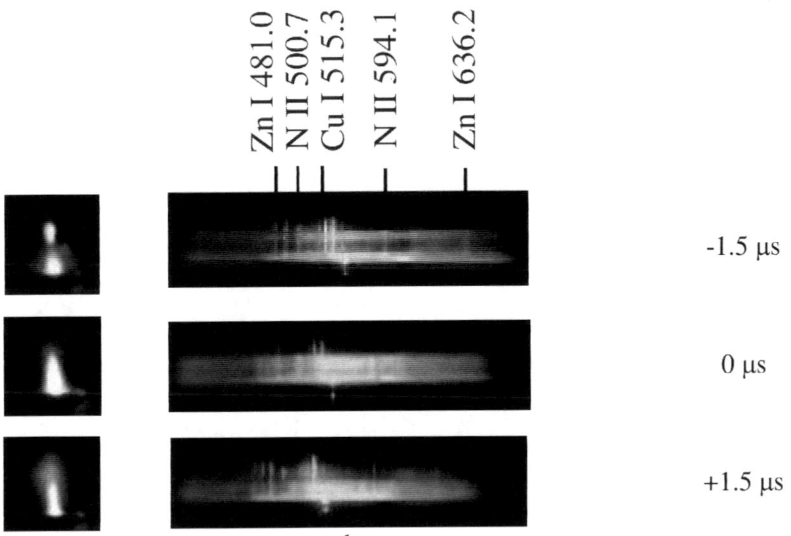

a *b*

Figure 1. Structure of laser plasma plume (a) and its radiation emission spectra (b) under laser irradiation of the brass in various irradiation modes.

The obtained spectra enabled us to determine the electron temperature T_e in the near-surface laser plasma from the relative intensities of spectral lines, and the density of electrons n_e from the Stark broadening of hydrogen and carbon spectral lines. The experiments were conducted at the LI power density of $q_{0.532} = q_{1.064} \leq 4\times10^9 W/cm^2$ in a laser spot of diameter d_n = 0.30 mm. It was found out that the exposure to shortwave LI stimulates the formation of an erosion-produced plasma while the infrared LI produces predominantly an air plasma. The experimental results showed that the parameters of laser-produced plasma on the brass and graphite targets depend on the ordering of laser pulses and the time interval between them. In the experiments with brass, the maxima of plasma temperature ($24\cdot10^3$ K) and electron density ($4.5\cdot10^{17}$ cm^3) are attained in the case of outrunning shortwave irradiation at λ = 532 nm, but at different values of the time interval between laser pulses, $\Delta t = -5$ μs and $\Delta t = -1.5$ μs, respectively.

In the experiments with graphite the spectral lines N II 399.5 nm and the group of lines N II (499.4 nm – 500.7 nm) were used to determine the plasma temperature (figure 2a), while electron density was evaluated from the measurements of the broadening of carbon spectral lines C II 426.7 nm (figure 2b). The maxima of plasma temperature ($18\cdot10^3$ K) and electron density ($4.7\cdot10^{17}$ cm^{-3}) are attained in the case of outrunning shortwave irradiation at λ = 532 nm, but for different time intervals between laser pulses, $\Delta t = -2.5$ μs and $\Delta t = -1.5$ μs, respectively. The differences of the values of inter-pulse time interval, which correspond to the maximum temperature and plasma density, can be explained by the complexity of the spatial structure of a dynamic plasma.

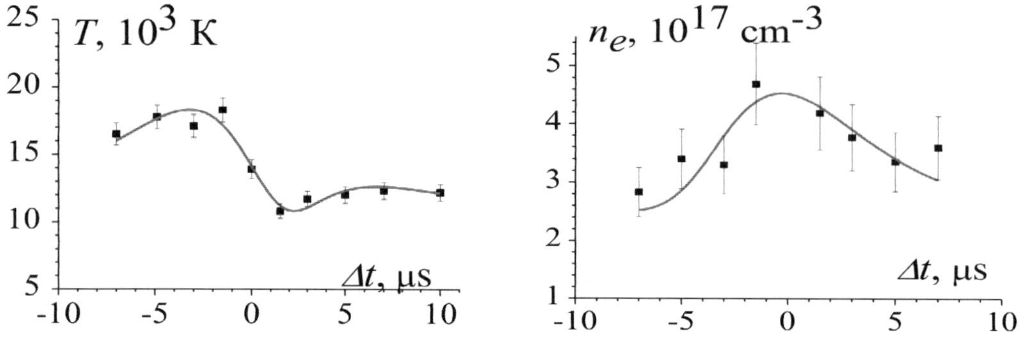

a *b*

Figure 2. Dependence of temperature (a) and electron density (b) in the carbon plasma on the time interval between laser pulses.

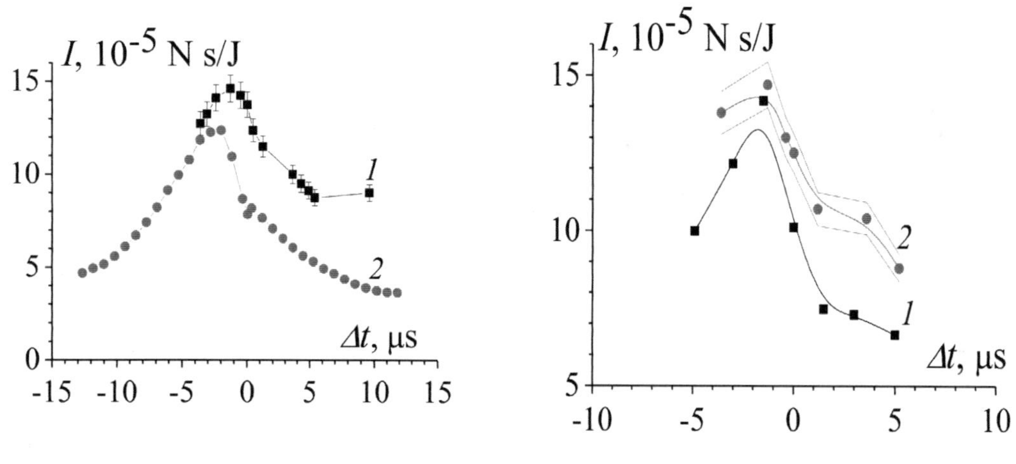

a *b*

Figure 3. Dependence of the measured (1) and calculated (2) recoil momentum on the time interval between laser pulses for the plasma on the brass (a) and graphite (b) targets.

This happens essentially due to a higher absorbing ability of the target and accumulation of erosive vapors under the influence of initial pulse with a shorter wavelength, as well as due to a quadratic dependence of the absorbing ability of plasma plume on a longer wavelength of the second acting laser pulse. Experimental dependencies of recoil momentum on the time interval between laser pulses are in satisfactory agreement with the estimates obtained using the equation of state for ideal gas and the value of laser pulse duration (figure 3a – for brass, figure 3b – for graphite). On the basis of obtained parameters and spatial structure of the plasma, it can be concluded that the most compact form of the plasma plume with the minimal expansion into the surrounding environment corresponds to maximum values of its parameters: plasma temperature, charged particle density and specific recoil momentum.

4. Conclusion

In this paper, we report on the dependencies of temperature, charged particle density and recoil momentum of a laser-produced plasma on the ordering of the double-pulse laser irradiation of the brass and graphite targets in air at wavelengths of 1064 and 532 nm and on the time interval between pulses. The characteristic size of radiation emission area of the plasma torch was determined using the photographic recording of a laser torch on a brass target in air for different modes of bichromatic laser irradiation. It was shown that the optimal conditions for recording the spectra of erosion-produced plasma of brass and graphite are provided in the case of outrunning shortwave laser irradiation at λ = 532 nm with the time interval between the pulses within 3-6 μs range. As for generation of the recoil momentum, the optimum time interval is 1.5-2 μs. These experimental results are essential for the enhancement of the sensitivity of laser spectral analysis and of the efficiency of laser plasma microengines.

References

[1] Minko L Ya, Chumakov A N and Bosak N A 1990 *Sov. J. Quantum Electronics* **17** 1480
[2] Minko L Ya, Chumakov A N and Bacanovich G I 1994 *Journal of Applied Spectroscopy* **61** 476
[3] Petuh M L, Rozantsev V A, Shirokanov A D and Jankowski A A 2000 *Journal of Applied Spectroscopy* **67** 798
[4] Pershin S M 2009 *Quantum Electronics* **39** 63
[5] Cremers D and Radziemski L 2009 *Laser-induced breakdown spectroscopy (Moscow: Technosphere)* 360
[6] Lubchenko F N, Fedenev A V, Bosak N A, Chumakov A N, Panchenko A N and Tarasenko V F 2009 *Cosmonautics and Rocket Engineering* **3** 62
[7] Chumakov A N, Avramenko V B and Bosak N A 2012 *Journal of Applied Spectroscopy* **79** 279-287

Electrode system configuration influence on micropinch discharge emission characteristics

E I Dodulad[1], **V A Kostyushin and O A Bashutin**
National Research Nuclear University MEPhI (MEPhI), Kashirskoe shosse 31, Moscow, 115409 Russian Federation

E-Mail: eidodulad@mephi.ru

Abstract. Article is concerned with electrode system geometry influence on high-current low-inductance vacuum spark discharge plasma development and its emissive characteristics.

1. Introduction

Pulse spark discharges are known to be intense sources of vacuum ultraviolet, x-ray radiation and charged particles flows [1]. In such discharges electrodes erosion products make major contribution in discharge plasma material, and they determine spectral composition of plasma radiation. In high-current low-inductance vacuum spark (HLVS) formation of hot dense plasma is possible due to pinching effect [2].

In previous works we started to investigate the influence of electrode system configuration on radiating plasma development in HLVS [3]. Those experiments were carried out on the same experimental setup with one discharge current and triggering circuit parameters. Herewith voltage polarity influence on each electrode material contribution in discharge plasma was studied.

For commonly used in such setups electrode configuration with pointy inner electrode and plane outer electrode sufficient dependence of elemental plasma composition on voltage polarity was found out. It should be noted that anode material radiation was observed in all cases.

Present article is concerned with electrode system geometry influence on HLVS discharge plasma development and its emissive characteristics.

2. Experimental setup

Experimental setup was fully described in our previous works [3]. Electrode system is presented on figure 1. In present paper inner electrode was made in form of a cylinder with flat base 20 mm in diameter. Outer electrode was in form of a needle with diameter 3 mm ("plane-point" configuration). Each electrode was made of different material (one of steel, the other of copper) in order to determine presence of their material in discharge plasma. Elemental composition of HLVS discharge plasma was studied by means of K_α radiation of each electrode material from interelectrode gap. Two focusing spectrographs with spatial resolution (FSPR-1D/2D) assembled according to Johann scheme were used for spectra registration.

[1] To whom any correspondence should be addressed.

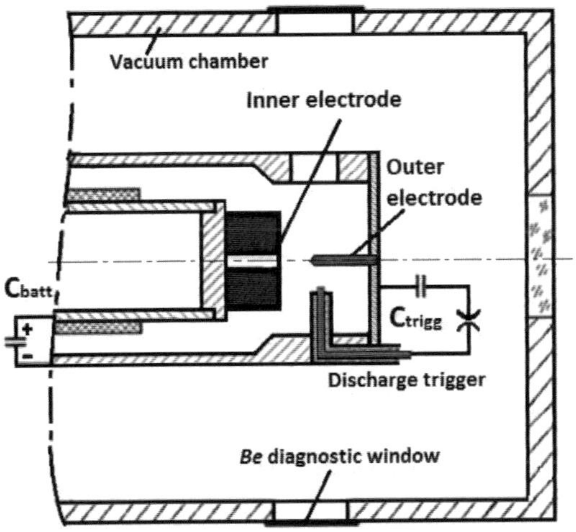

Figure 1. Electrode system of the experimental setup.

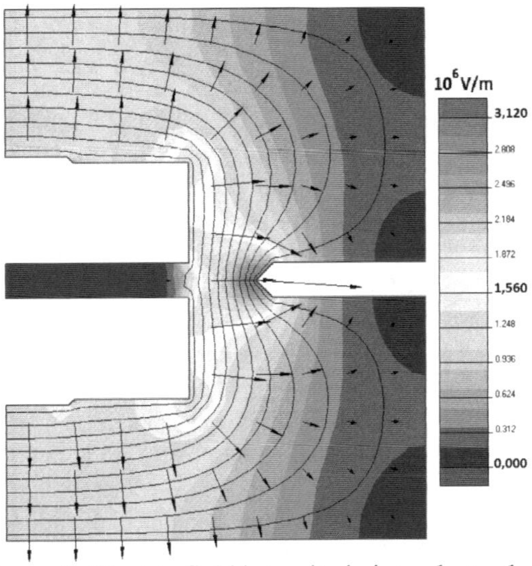

Figure 2. Electric field intensity in interelectrode gap of experimental setup.

Spectrographs were placed on two sides of vacuum chamber and either was adjusted to material of one of electrodes. At the same time pinhole images of the discharge in x-ray range were obtained (figure 3) [3].

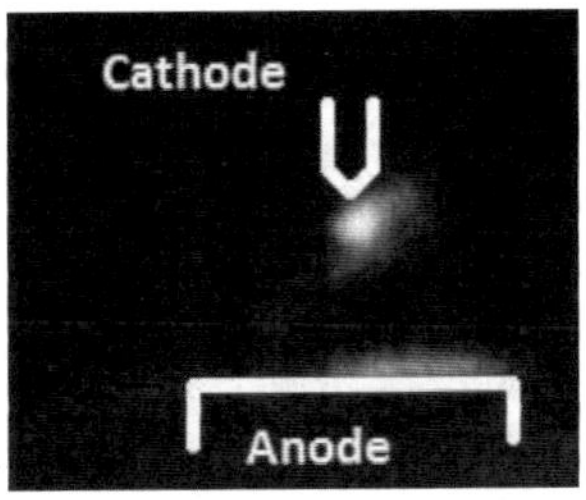

Figure 3. X-ray pinhole image of HLVS discharge.

Spectra of K_α radiation obtained and corresponding densitograms are represented on figure 4.

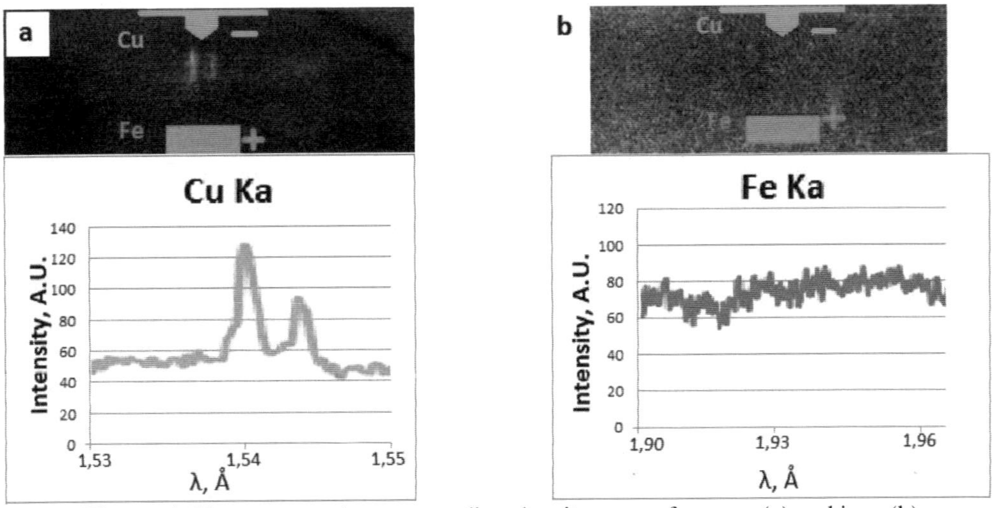

Figure 4. K_α spectra and corresponding densitogram of copper (a) and iron (b).

3. Discussion

Measurements for "plane-point" configuration were made only with inner plane electrode acting as anode. In case of reverse polarity, dishonorable discharge developed between the edge of the plane electrode and outer current conductor. This phenomenon was confirmed by x-ray pinhole imaging. The disruption was caused by high electric field intensity near the edge of plane cathode and the proximity of the outer current conductor.

$K_{\alpha 1}$ and $K_{\alpha 2}$ spectral lines of pointy copper cathode's material are distinguished on figure 4a. They are radiated from the area 2 mm in length and 1 mm distant from cathode. In the same time K_α radiation of erosion products of inner anode (iron) are practically non-existent (figure 4b).

Absence of plane anode erosion products radiation can be explained in the following way. In this case, electric field intensity near the plane inner electrode increases from its center to its edges. Due to this initial plasma expands in interelectrode gap from cathode to anode by a wide cone. In this case anode area, bombarded by electron beam, grows by several times. Power density deposited on the anode surface diminishes respectively and so does the surface heating. Since metal vapor emission occurs mostly in direction normal to electrode surface, with distance from the discharge axis the beam intersects smaller area of anode material vapor. As a result ionization rate of vaporized material decreases as well as extra influx of electrons from anode region and secondary electrons from cathode. For this reason inflow of erosion products in discharge gap decreases sufficiently.

One can notice the shift of pinching area from discharge axis on the most of pinhole images of the discharge in "plane-point" configuration. It can be seen that this shift occurs in the direction of maximum of electrical field intensity, which is evidence of initial electrical field distribution influence on the following plasma distribution and emission characteristics of the discharge.

4. Conclusions

- Unlike configurations with pointy anode, in case of configuration with plane anode K_α radiation of anode material is not observed. Its absence can be explained by the influence of electric field intensity distribution.
- K_α spectral lines of pointy cathode's material are radiated from the area 2 mm in length and 1 mm distant from cathode.
- Experiments provide evidence of initial electrical field distribution influence on the following plasma distribution and, as a result, on emission characteristics of HLVS discharge.

References

[1] Papernyi V L and Lebedev N V 2014 *Plasma Physics Reports* **40** 78-82
[2] Sarantsev S A 2015 *Physics Procedia* **71** 165-170
[3] Bashutin O A, Vovchenko E D, Dodulad E I, Savjolov A S and Sarantsev S A 2012 *Plasma Physics Reports* **38** 235-243
[4] Bashutin O A, Vovchenko E D, Savjolov A S and Sarantsev S A 2010 *Problems of Atomic Science and Technology* **6** 168-170

Journal of Physics: Conference Series **666** (2016) 012014 doi:10.1088/1742-6596/666/1/012014

Method of measuring the time of x-rays transfer in the closed cavities and in the mode of thermal breakdown

N M Romashkina[1], A P Muntyan, S I Petrov and S S Taran
RFNC-VNIIEF, pr. Mira, 37, Sarov, 607188 Russian Federation

E-mail: dushinala@gmail

Abstract. Method for the study of transfer processes of X-ray radiation (XR) in closed cavities and in thermal breakdown mode is developed and described. Experiments are carried out on "Iskra-5" facility using two RFR-4 streak cameras.

1. Introduction

The presented method is designed for the study of transfer processes of X-ray radiation (XR) in closed cavities and in thermal breakdown mode occurring during interaction of high-power laser radiation of the second harmonic of iodine laser "Iskra-5" with X-ray converter boxes.

The X-ray source is a cylindrical target-converter of "Porthole" type [1], irradiating or a hollow box with a slit on the side through which the rate of propagation along the box is measured, or a set of foils of different materials. The delay time of occurrence of radiation behind the foil and is the time of thermal breakdown.

2. Experimental setup

For measurements the two perfectly calibrated RFR-4 X-ray streak cameras and SKHR7 X-ray time frame magnifier are used. First RFR-4 observes the side wall of the target through the laser radiation input hole and is designed to define the parameters of XR, illuminating the investigated targets. Second RFR-4 [2] and SKHR7 both observe either the gap in the side surface of the additional target-channel or the studied foils, detectors are designed to measure the XR transfer time. To ensure the spatial resolution along the slit and behind different foil-filters the RIVS-4 spectrographs [3] are used. In the streak cameras the measurements are conducted in the spectral range from 0.2 keV to 1.1 keV in 5 narrow intervals with the width values hv/dhv~3-7, with spatial resolution ~150 microns and temporal resolution ~40 ps. The measurement margin of the absolute XR flows is ~20%. Frame magnifier registers 11 target images with a spatial resolution ~30 microns, 100 ps delay between frames and exposure time of each frame 100 ps [4].

In the experiments described, the Planck radiation temperature in the "Porthole" amounted to 110-150 eV, the rate of radiation propagation in the closed cavities was in the range 1-12 mm/ns, while the thermal breakdown of the foils of different materials was in the range of 50-550 ps.

References

[1] Lazarchuk V P, Muntyan A N, Murugov V M, Petrov S I and Senik A V 2008 *Instr. and Exp.*

[1] To whom any correspondence should be addressed.

Techniques **51** 232-236

[2] Lazarchuk V P, Litvin D N, Murugov V M, Petrov S I, Pryanishnikov I G and Senik A V 2004 *Pribory I Tekhnika Eksperimenta* **2** 128-132

[3] Lazarchuk V P, Muntyan A N, Murugov V M, Petrov S I and Senik A V 2006 *Plasma Phys. Rep.* **32** 129-134

[4] Lazarchuk V P, Muntyan A N, Murugov V M, Petrov S I and Senik A V 2004 *Instr. and Exp. Techniques* **47** 255-259

Calibration source based on capillary discharge

E Baronova[1] and A Stepanenko

National Research Centre "Kurchatov Institute", Kurchatov square, 1, Moscow, 123182, Russia

E-mail: baronova04@mail.ru

Abstract. A source of ultraviolet radiation based on capillary discharge was made. Alignment and calibration of focusing spectrograph using this source is described.

Capillary discharge is one of the simplest type of Z-pinch discharge. It is thoroughly studied due to its lithography applications. The modes with minimal scattering of ceramics and electrodes materials as well as with maximal output of shortwave radiation (e.g. radiation in the vicinity of 135 Å for which Mo-Si mirrors are used) are of particular interest. These modes are achievable in capillaries filled with Xe at relatively small current which would be sufficient for generation of radiation in the vicinity of 135 Å. It is necessary to note that in capillary discharges filled with Xe radiation with wavelengths 110-120 Å is the more intensive than that with 135 Å. However, mirrors for 110 Å contain Be, forbidden in industrial facilities.

In this work we obtained the spectra (100-700 Å range) of the capillary discharge, which was used for the alignment and calibration of UV spectrograph. Focusing spectrographs with small incidence angles are very sensitive to alignment, i.e. localization of a precise positions of the detector and the dispersive element. Capillary discharge is chosen for alignment procedure since it is cheap, reliable (capillary withstands tens and hundreds of thousands of shots without replacement of its parts) and easy to operate. Along with that, the radiation source is fixed in space and small, and its radiation is intensive.

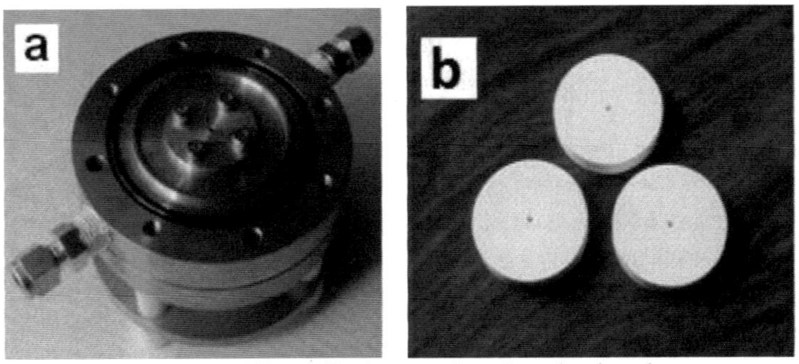

Figure 1. a – capillary discharge chamber, b - insulator.

[1] To whom any correspondence should be addressed.

Figure 1a shows a capillary discharge chamber. The capillary is 1 ÷ 2 mm in diameter, the length of cylindrical channel of the capillary is 6 mm. The capillary is designed as a small channel inside a ceramic disk 6 mm in thickness and 30 mm in diameter. The voltage within 3 ÷ 10 kV range is applied to two ring-shaped electrodes. The high negative voltage is applied to the top flange while the bottom flange is grounded. The outlet flange of capillary has a cylindrical aperture for collimation of radiation. The exterior surface of the flange has a leak-proof mounting unit for the filter. Figure 1b shows the photographs of ceramics used in the capillary chamber.

In our experiments the working) gas (Xe) was puffed into the capillary through side tubes. Turbomolecular pumps were used to achieve high vacuum. The pressure of Xe was of 0.2÷0.5 Torr. Figure 2 shows the experimentally obtained current waveform. The maximum current was 3.7 kA, half-period of the current was 1200-1300 ns.

Figure 3 shows the spectrograph used for spectrum registration. The spectrograph was equipped with spherical holographic grating (30mm x 40mm, 4200 lines per mm) covered with gold, grazing angle was 10°. The measurements were conducted in the 1st diffraction order, DEF film was used as a detector.

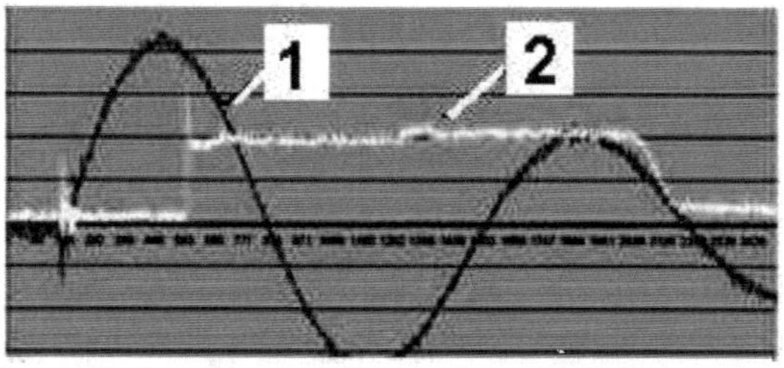

Figure 2. 1 – shape of the current, 2 – voltage shape.

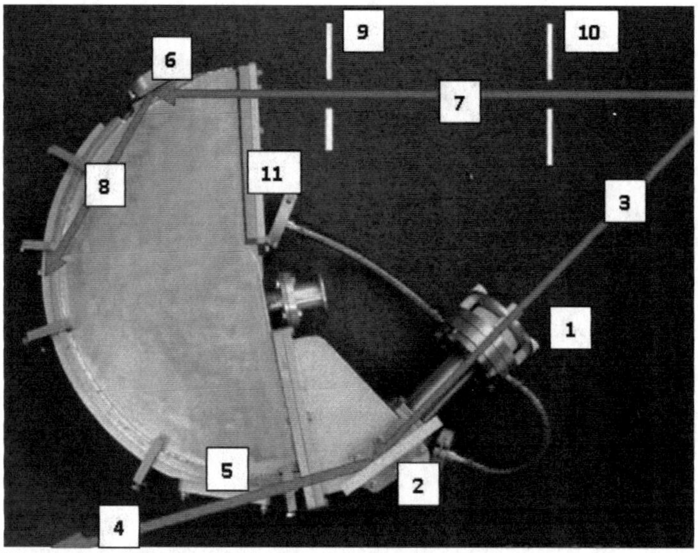

Figure 3. 1 – capillary chamber, 2 - grating, 3 – incoming laser beam for alignment of EUV, UV channel, 4 – reda laser beam reflected from grating, 5 – position of detector (film, CCD,MCP, etc.) 6 – grating position , 7 – incoming laser beam for alignment of second EUV,UV channel, 8 – reflected laser beam, 9, 10 - diaphragm, 11 – second input flange for EUV,UV radiation.

Width (1.5 mm) and height (7 mm) of the slit were determined by sensitivity of spectral channel. A 250 Å thick carbon filter deposited on the wire grid (cell size of 250 micron, wire diameter of 15 micron) was installed on the slit. The filter is transparent for the visible radiation, so it was possible to perform initial optical alignment of the spectrograph by red laser.

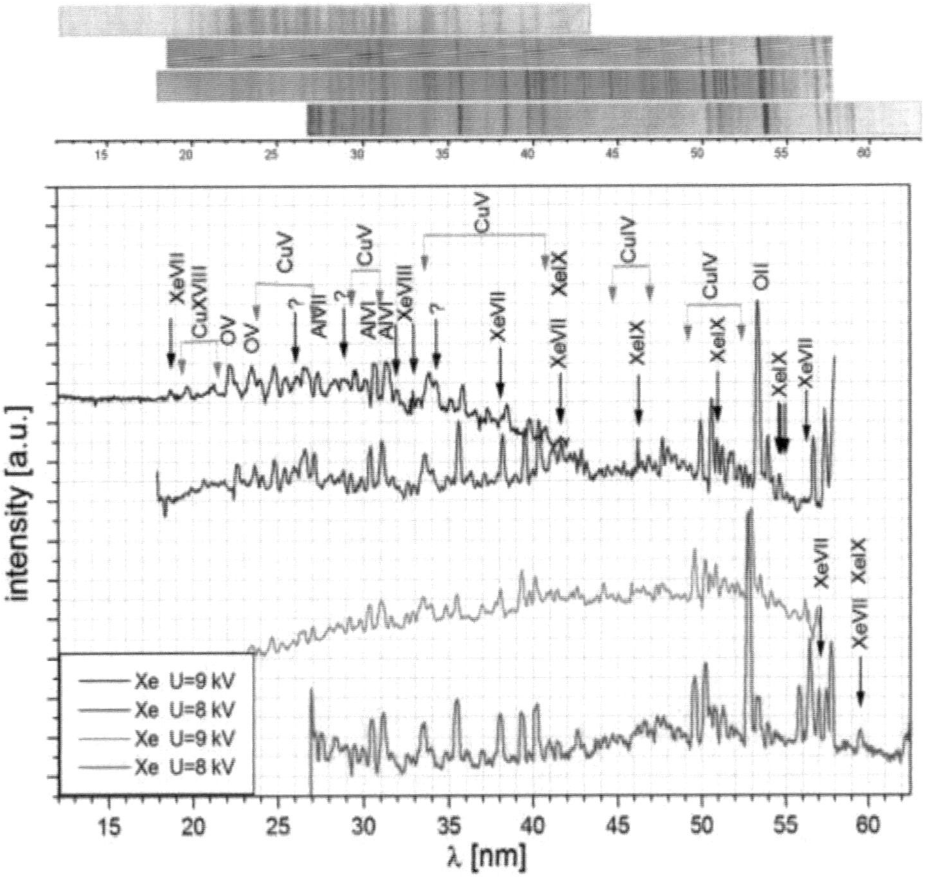

Figure 4. Spectrum registered from capillary discharge and the corresponding densitogram.

Reflection coefficients of mirrors and dispersive elements for EUV, UV radiation at large grazing angles are pretty low, so it forces us to work at small grazing angles. This makes alignment of the device more complicated. At the first stage of the alignment detector and dispersive elements are put in calculated positions (so called geometrical alignment) and then an optical alignment by checking precise positions of elements using visible light sources (red lasers) takes place. The last stage is the final adjustment using EUV, UV calibration source (capillary).

Spectrum of calibration source within given wavelength range is registered by spectrograph with further evaluation of half-widths and relative intensities of lines and repositioning of spectrograph units for achieving optimal parameters. Optimally focused spectrum obtained by 4200 lines per mm grating and corresponding densitograms is shown in figure 4. The spectrum contains lines of ions of Al, Cu, O and Xe with various degrees of ionization. Registered spectra shows_that the contribution of lines of electrode material to plasma radiation must be taken into account when simulating the processes in capillary discharges.

Thus, the capillary discharge was created, and its spectra including lines of working gas, electrode material and ceramics were registered. The designed source was successfully applied for alignment and calibration of focusing spectrograph. Presence of intensive spectrum containing lines of electrode (cathode) material testifies that it is reasonable to deposit the corresponding material on the inner surface of electrode to obtain higher radiation yield within the desired wavelengths.

Comprehensive optical diagnostics of laser-induced plasma objects

V K Goncharov, K V Kozadaev[1], H P Mikitchuk and A G Novikau
Belarusian State University, Kurchatova Street 7, Minsk, 220138 the Republic of Belarus

E-mail: kozadaeff@mail.ru

Abstract. A new research facility for comprehensive analysis of laser-induced plasma objects equipped with high resolution devices for space, time and spectral measurements was developed. The results of combined use of laser probing and spectral analysis of metal erosion laser jets (ELJ) allowed us to propose and justify models of laser erosion of metals for various modes of exposure.

1. Introduction

Monitoring of energetic, temporal, spatial and spectral characteristics of plasma objects arising as a result of laser exposure is of great practical interest for developing high-precision laser technologies of material processing [1-3]. The main goal of this work was development of an automated hardware and software research complex equipped with high resolution devices for precision measurements in space, time and spectral domains, which enables optical diagnostics of laser-induced plasma formations.

2. Research complex for diagnostics of laser-induced plasma objects

The schematic diagram of the suggested research facility shown in figure 1 is comprised of 5 units: (I) is the unit where an acting pulse is formed and its characteristics are registered, (II) is for studying the spectral characteristics of (ELJ), (III) is for laser probing of ELJ, (IV) is for experimental data processing, and (V) is the synchronization arrangement.

Unit (I) incorporates laser (1), deflecting plates (2) and (3), pulse form recorder (4) (photo diode) and calorimeter (5).

Unit (II) is designed for high spectral and spatial resolution investigation of ELJ [4] using approaches of Laser-Induced Plasma Spectroscopy (LIPS).

[1] To whom any correspondence should be addressed.

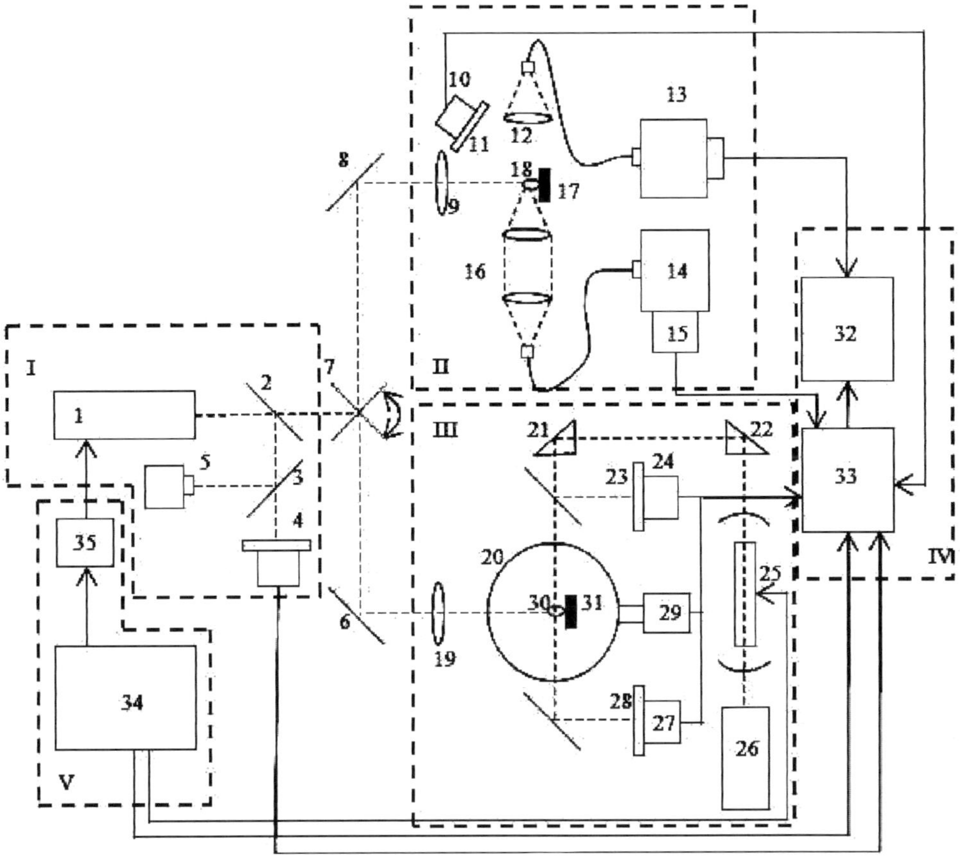

Figure 1. Diagram of the facility for optical investigation of laser-induced plasma objects. 1 – exposing laser; 2, 3 transparent deflecting plates; 4 – pulse form recorder; 5 – pulse energy recorder; 6 – 8 – mirrors; 9, 19 – exposing radiation focusing lenses; 10 – ELP integral glow censor; 11, 23, 28 – optical filters; 12, 16 – optical systems for ELP image formation at the entrances of the spectral devices; 13 – spectrophotometer, 14 - monochromator, 15, 29 – photoelectric multiplier; 17, 31 – laser targets, 18, 30 – ELP.; 20 – integrating sphere; 21, 22 – rotary prisms; 24, 27 – probing radiation intensity detectors; 25 – probing laser, 26 – auxiliary alignment laser; 32 – portable computer; 33 - multichannel ADC; 34 – sync pulses generator; 35 – exposing laser synchronization optocoupler.

Using LIPS along with Time Resolved Laser Spectroscopy (TRLS) makes it possible to study the temporal structure of glow spectrum of plasma objects [5], which, in its turn, allows us to have a better picture of physical processes taking place in ELJ of metals as a result of exposure to laser irradiation. Investigation of dynamics of ELP integral glow can also be of practical interest [6].

To accomplish the aforementioned research tasks the facility includes spectrophotometer (MS 2001i) with a CCD acting as a light detector (13); monochromator (14) coupled with the silicon photomultiplier (SPM 10020 SensL) and photodiode (10) equipped with an interference mirror (11) (λ=1064 nm) to neutralize the effects of exposing radiation. In addition to that, the unit incorporates a pointing system (7-9) which directs the radiation at the surface of the target (17), together with optical systems (12) and (16) for imaging of ELJ (18) at the entry points of spectral devices.

Unit (III) uses laser probing of ELJ [4, 7] to examine the stages of formation and evolution of ELP. This unit contains a system for forming and targeting of probing pulse at ELJ (21, 22, 25, 26), a targeting system (6, 7, 19), which allows to direct the exposing pulse at the target (31), as well as a system for registering scattered and absorbed components of probing radiation (23, 24, 27, 28, 29). It should be noted that in this case the light detectors are shielded with interference filters which protect them from spurious radiation and minor flares.

Data processing unit (IV) consists of a multi-channel 100 MHz digital oscilloscope 33 (B422) and a portable computer 32. The software bundle developed for the several basic modes of the facility allows automated recording, initial processing and representation of obtained experimental data.

Unit (V) includes the G200P multichannel generator of delayed pulses (34), which provides separate triggering pulses for each channel within the wide range of their durations and delays. Besides, optocoupler (35) eliminates accidental triggering of exposing laser by induced interference.

3. Summary

The optical measuring stand described in this paper provides a comprehensive study of the erosion of ELJ that occur when metal targets are exposed to high-power nanosecond laser pulses. The obtained experimental results make it possible to determine the main regularities in the formation and kinetics of liquid-drop phase of metals in near-surface region of the target under atmospheric pressure. Condensation mechanism of formation of liquid drops, originated in nanosecond laser-induced erosion torches was experimentally found and justified. The results of these studies formed the basis for new methods of metallic nanostructures formation for various purposes.

References

[1] Mirkin L I 1975 *Fizicheskie osnovy obrabotki materialov luchami lazera (Physical Foundations of Material Processing by Laser Beams)* (Moscow: MSU)
[2] Rykalin N N, Uglov A A, and Kokora A N 1975 *Lazernaya obrabotka materialov (Laser Processing of Materials)* (Moscow: Mashinostroenie)
[3] Anisimov S I, Imas Ya A, Romanov, G S and Khodyko Yu V 1970 *Deistvie lazernogo izlucheniya bol'shoi moshchnosti na metally (Action of High Power Laser Radiation on Metals)*, (Moscow: Nauka)
[4] Goncharov V K , Kontsevoi V L, Puzyrev M V and Smetannikov, A S 1995 *Instr. and Exp. Techn.* **38** 661
[5] Demtroder W 2008 *Laser spectroscopy: Vol. 2 Experimental Techniques* (Berlin: Heidelberg)
[6] Goncharov V K, Kozadaev K V and Shchegrikovich D V 2013 *J. Appl. Spectr.* **80** 395
[7] Goncharov V K, Kozadaev K V and Shchegrikovich D V 2013 *J. Eng. Phys. Thermophys.* **86** 798

Journal of Physics: Conference Series **666** (2016) 012017 doi:10.1088/1742-6596/666/1/012017

Quadrature laser interferometry in the pulsed plasma diagnostic

A P Kuznetsov[1]
National Research Nuclear University MEPhI, Kashirskoe shosse 31, Moscow, 115409, Russian Federation

E-mail: apkuznetsov@mephi.ru

Abstract. The article provides an overview of methods for plasma diagnostics based on the laser interferometry with quadrature recording of informative signals. Quadrature interferometers allow providing a wide dynamic range for measurements of electron densities while maintaining the homogeneous differential sensitivity. The existing range of linear electron density measuring with quadrature interferometers in pulsed plasma installations equals to 10^{10}-10^{19} cm^2.

1. Introduction

One of the most informative parameters in describing the state of plasma is its electron density. The most precise information about the electron density in plasmas can be obtained from measurements of phase perturbations in the probing wave by laser interferometry. The phase shift, δ, in the probing electromagnetic wave propagated in plasma over a distance ℓ is determined by the change of the plasma refractive index n that is integral over the probing chord:

$$\delta = \frac{2\pi}{\lambda}\int_0^{\ell}(n-n_0)d\ell \qquad (1)$$

where λ is the wavelength of the probing radiation. The refractive index n in the optical wavelength range of the probing radiation at the degree of plasma ionization exceeding 1÷10% is mainly determined by the electron density N_e. Therefore, in the absence of magnetic field and with the neglection of collisions, the refractive index can be approximately written in the following form:

$$n \approx 1 - 4.49\times10^{-14}\lambda^2 N_e \qquad (2)$$

Interferometry methods allow measuring the change of the refractive index, $\Delta n=(n-n_0)$, and, respectively, to determine the value of $\int_0^{\ell} N_e d\ell$, which is called the linear electron density.

Interferometers being used in the plasma diagnostics can be classified by three parameters:

- the laser working mode: *conventional* and *autodyne*;

[1] To whom any correspondence should be addressed.

- the interference signal registration method: *interferometers with interference field visualization* [1] and *interferometers with photoelectric mixing* [2];
- the number of rays involved in the interference: *double-beam* [3] and *multibeam* [4].

A vast number of practical implementations of interferometers in plasma diagnostics are based on the schemes in which the plasma introduces phase changes in the probing electromagnetic wave without affecting the radiation source itself. The optical schemes for interferometers in which a laser is used only as an illuminator can be called the conventional ones. Methods of the autodyne laser interferometry are based on the dependence of the power and frequency of laser radiation on the change of the refractive index of the researched plasma placed into a laser cavity [5], or into an additional resonator optically coupled to the laser [6].

Signal processing using the photoelectric registration can be carried out both without the transfer of the signal spectrum in the homodyne schemes and at the intermediate frequency in the heterodyne schemes. For more informative and interference immune measurements, multiple phase-correlated channels for photoelectric detection can be used (differential [7], quadrature [8], quadrature-differential (push-pull) [9]).

The most versatile, possessing the ability to be integrated into a wide class of plasma installations are the measuring systems based on the techniques of interferometry with the quadrature photodetection. The quadrature detection of interference signals and the phase recovery procedures are the methods of the orthogonal correlative processing of the signals of the maximum likelihood. Realization of the quadrature algorithms is universal and provides interferometers with a wide dynamic measuring range while maintaining a homogeneous differential sensitivity. The sensitivity and the accuracy of interferometer measurements are limited by photographic noise. Temporal resolution essentially limited by the transit time of the probe radiation through the plasma can practically be determined by the speed of photorecording devices.

2. Interferometry with quadrature photoregistration of signals

For the first time in the optical diagnostics of plasma, a quadrature scheme was applied in the middle of the 70s of the XXth century by Buchenauer and Jacobson [8]. In the microwave plasma diagnostics, this technique had been applied much earlier, and is known as the "interferometer with polar indication" [10]. However, due to the complexity of the analog implementation of signal processing procedure, it has not become widespread for a long time.

The basic idea of the quadrature interferometer involves the use of two interference signals shifted relatively to each other by a quarter period. The output of the interferometer is formed by the two types of signals:

$$U_1(t) = U_0(t)\sin(\delta(t)) \tag{3}$$

$$U_2(t) = U_0(t)\cos(\delta(t)) \tag{4}$$

where $U_0(t)$ includes the multiplicative noise related to the amplitude fluctuations of the interfering fields. Transformation of these signals, similarly to the transformation of the rectangular coordinate system into the polar one, allows restoring the optical length variation:

$$\delta(t) = \text{arctg}\left(U1(t)/U2(t)\right) \tag{5}$$

The signal division operation removes the influence of the slowly changing multiplicative noise $U_0(t)$ in the interference signal in comparison to the functions $\sin(\delta(t))$ and $\cos(\delta(t))$.

The quadrature interferometer allows precise registration of almost all phase incursions with high homogeneous differential sensitivity $\mathrm{d}U/\mathrm{d}\delta$.

Nowadays, the most effective realization of the quadrature signals processing algorithm is possible with the use of digital systems for data collection followed by its computer processing. It should be noted, that the transformation (5) allows the precise measurement of the phase shift only in the range of main values of arctg function limited to one phase cycle $-p/2 < \delta \leq p/2$. Therefore, if the optical length $n\ell$ change occurs during the measurement process for a value exceeding $\lambda/2$, the current value of $\mathrm{arctg}(U_1/U_2)$ jumps on a π losing all previous phase data (figure 1d). The regular algorithm of "phase expansion", i.e. the phase values transformation for intervals [-π/2, π/2] into a continuous phase function, consists of the detection of phase values fluctuations between two nearby points, exceeding π/2, and by addition or subtraction of π rad phase jumps. When the interference signal with phase fluctuations within the phase cycle is computer-processed, the operational margins of the calculations increase, thus the more noise-resistant algorithms are needed to recover the continuous phase function. [11, 12].

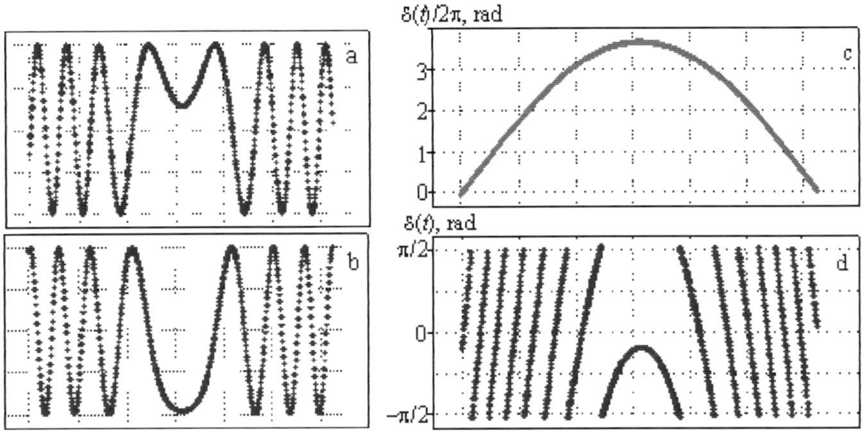

Figure 1. Quadrature interference signals (a, b) and the phase "expansion" recovery result (c) with parabolic changing of the medium refraction value (d).

3. Optical methods of quadrature signals forming in homodyne interferometry

Optical methods of forming the measurement channels in the quadrature homodyne interferometry are the most versatile and can be used for measurements not limited by the amount of phase shift values. Thus quadrature signals are formed in spatially separated channels and detected by several detectors. The quadrature signals may be formed in various ways, for example by using two photodetectors placed at different regions of the interference fringe with a distance equal to a quarter of the fringe. This is achieved by moving the diaphragm detectors across the direction of the interference fringe. The current phase difference value of the photodetector signals during adjustment can be monitored at a two-channel oscilloscope in the scanning mode y(x) (figure 2). If the setup is adjusted to provide the modulation of the wavelength of one arm of the interferometer by an amount >λ/2, an ellipse will generally be observed on the oscilloscope screen. In the case of equal amplitude signals from the detectors the quadrature mode corresponds to a circle.

When the interferometer is configured to the mode of infinitely wide fringe, the quadrature signals can be generated using a quarter-wave plate mounted at 45 degrees to the plane of polarization of the probing radiation and orthogonal polarizers [13] (figure 3). Quarter-wave plate transforms the polarization of the radiation in the reference arm from linear to circular, that can be represented as a superposition of two linearly and orthogonally polarized components which are phase-shifted by π/2. As a result of mixing the measuring and reference waves, the projections of circular polarization waves interfere with the linearly polarized measurement wave on the mirror M_4. The interference signals, which are orthogonally polarized and quadrature to each other, are spatially split by the polarizing prism (PBS) to enter the measuring channels. Due to the fact that usually a pulsed plasma discharge produces electromagnetic interference, the registration part of the interferometer is moved outside the experimental hall. To this end, the radiation is focused on the ends of the fiber optic cable contained in the XY adjustable mountings, (D_1, D_2), and transported to the input of the photodetector module. After photoelectric conversion the informative signals are recorded with a digital oscilloscope and transferred to a computer for further processing.

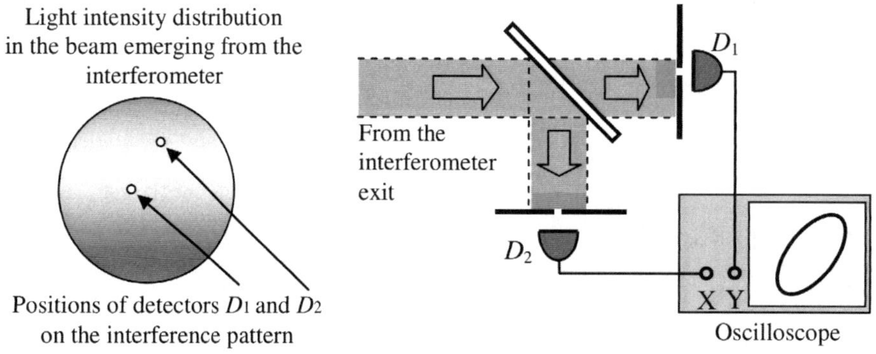

Figure 2. Principle of forming a quadrature recording channel.

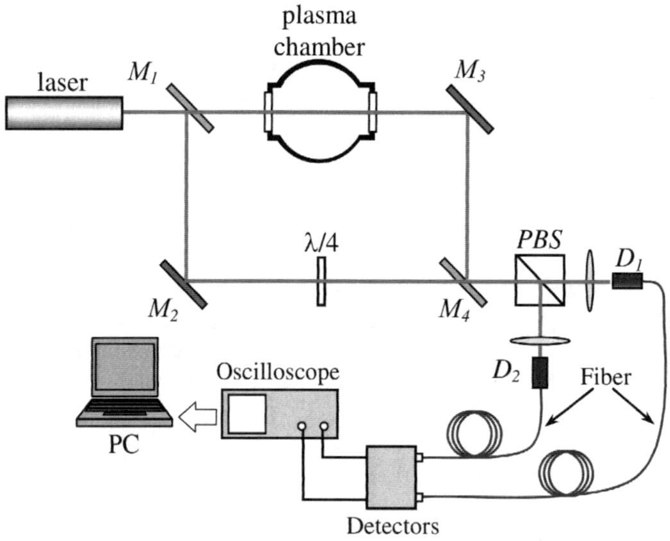

Figure 3. Schematic of the quadrature interferometer: (M_1–M_4) mirrors of the Mach–Zehnder interferometer.

Our experimental studies have shown that the quadrature homodyne interferometer allows achieving the accuracy of measurement of the phase shift at the level of $(3\cdot10^{-7})$ rad*Hz$^{-1/2}$.

Easiness of implementation, high homogeneous differential sensitivity and large measurement range allow integrating the quadrature interferometers in a class of plasma installations. Thus, different implementations of the homodyne quadrature interferometer have been developed by us for a research of temporal dynamics of the spatial distribution of the plasma electron density [13] and the gas-dynamic pressure of corpuscular streams [14,15] in the micropinch discharge of the low-inductive vacuum spark type, to measure the dynamics of the electron density of the erosion capillary discharge in air at atmospheric pressure [16], the electron density and the degree of plasma ionization in the plasma target on the basis of a linear electric discharge in hydrogen [17].

The results obtained in various plasma installations show that quadrature homodyne interferometers allow measuring a wide range of plasma electron densities (10^{13}–10^{19} cm^{-3}) with a time resolution of ~ 10^{-9}. They maintain alignment in experimental hall conditions at an elevated level of mechanical vibrations and allow performing the measurements in the presence of high levels of electromagnetic interference, characteristic to the pulsed electrophysical installations.

4. Methods of quadrature signals forming in heterodyne interferometry

The functioning of a heterodyne interferometer is based on the photoelectric detection of the interference field of two coherent waves - reference and measurement - with different optical frequencies. The frequency shift *f* between the interfering beams can be produced in various ways: by the Doppler shift due to the reflection of radiation from a moving reflector [18], diffraction grating rotation [19], mixing radiation of two frequency-stabilized lasers [20], using dual-frequency lasers [36, 37] or by using acousto-optical cells [21]. The typical heterodyne interferometer scheme is shown in Figure 4a. The acousto-optic Bragg cell in the scheme performs two functions: it splits the laser beam into the reference and measurement beams and provides a frequency shift of the diffracted reference beam relative to the frequency of the measurement beam.

Unlike the homodyne interferometry, where information on the measured value of the change in the optical length *nL* is contained in the amplitude of the photoelectric signal, the heterodyne methods imply the reduction of measurement data processing to the high-precision phasemetry. Full implementation of the advantage of heterodyne interferometry is possible only under the condition that the phase detector, included in the registration system, provides a wide frequency band of signal registration, high sensitivity and a large dynamic range > 10^6. The conventional schemes of phase detectors are based on the precision measurement of the duration of each period of the sine carrier [22]. Phase detectors based on the measurement of time intervals are usually limited to the bandwidth of the measured signal <100 kHz. For this reason, in plasma diagnostics the heterodyne interferometers are traditionally used for the research concerning the slowly varying time-density plasma that is mostly distinctive for systems with magnetic confinement [19, 20]. Application of heterodyne interferometers in the diagnosis of a pulsed plasma with the lifetime of ~10^{-6} s and less requires the development of broadband phase detectors.

In studies of plasma in the volume pulsed-periodic discharges in air at atmospheric pressure, the parameters at which the discharge is ignited can vary widely [23]. To measure the temporal dynamics and spatial distribution of the electron density of the plasma in volume pulsed-periodic discharge in a gas flow in the chamber section of ~1m, we have developed a double-wave heterodyne interferometer which allows measuring the electron density in the range of 10^{10}-10^{13} cm^3 with a time resolution not worse than 10^8. The minimum detectable phase shift was 10^{-5} rad in the band of 1 MHz, which limits the minimum measured linear electron density of the plasma at $2*10^{10}$ cm^2 when using a CO_2 laser. To maintain such parameters, a scheme for direct quadrature analog-to-digital conversion of the phase-modulated signal, followed by digital processing of informative samples, was implemented.

The functional diagram of the analog-to-digital processing of the measuring signals of the heterodyne interferometer is shown in figure 4b. The clock frequency of synchronization of the whole circuit is set by a crystal oscillator (G) at a frequency f = 144 MHz. After dividing the frequency $f/4$ = 36MHz and preamplification, the control signal is sent to an acousto-optic modulator (AOM). The phase-modulated signal $A(t)=A_0(t)\cos(2pf_o t+d(t))$ from a detector (D) after the pre-amp goes to the input of ADC (16 bits) operating at a sampling frequency f = 144 MHz . Since the ADC is clocked at 4 times the frequency of the phase-modulated signal, this allows taking 4 samples per one period of the phase-modulated signal with a strictly fixed phase difference i–(i+1) ($\pi/2$), i–(i+2) (π), i–(i+3) ($3\pi/2$). Since the time intervals of sampling are generated from a reference frequency, the algorithm of a digital quadrature differential phase difference measurement method is implemented. Besides, measurement of 4 samples per one period of the phase-modulated signal eliminates problems associated with the ADC zero drift and allows lowering the noise level at even harmonics of the carrier frequency.

The digital path of signal processing is implemented on the basis of the programmable logic integral circuit (PLIC) and consists of multipliers (1), differential amplifiers (2) performing the summation of counterphased signals, algebraic module CORDIC (3) (COordinate Rotation DIgital Computer), based on iterative bit calculations of the *arctg* value (calculation accuracy of 10^{-7} rad). A software module for calculating the density of the plasma (4) is implemented into the PLIC. The calculated data on the value of the linear electron density N_e are accumulated in the RAM unit, and then are transmitted, under control of the microcontroller, through the LAN interface to a program processing the measurement results in the PC.

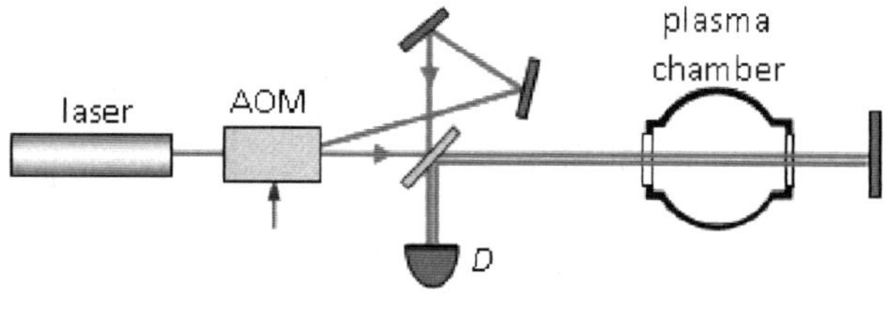

a)

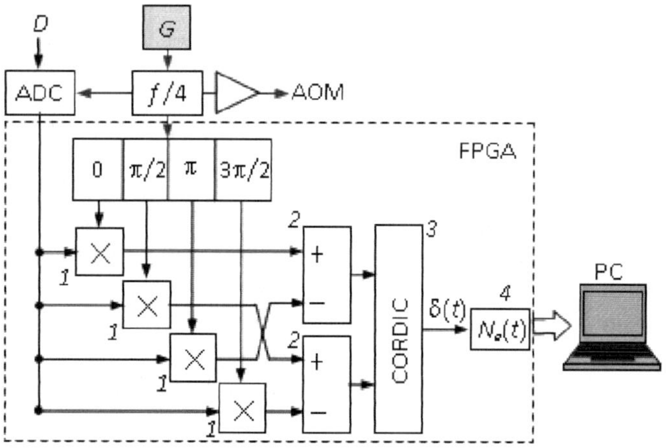

b)

Figure 4. Interferometer optical scheme (a) and informative signal processing functional scheme (b).

5. Methods of quadrature signals forming in autodyne interferometry

Almost simultaneously with the advent of lasers, it was noted that the external radiation reaching the resonator influences the mode of its generation. So since the early 60s there were proposals for the use of the method of autodyne (intracavity) admission to study the dynamics of the electron density of the plasma, placed in the way of the reflected radiation entering the laser [6, 24]. It should be noted that in plasma diagnostics the name "laser interferometers" was used exactly for autodyne interferometers [25]. The principle of operation of such laser systems is based on the autodyne effect, well-known in the radiowave and microwave ranges and inherent to all oscillatory systems. Essentially, the laser functions as a coherent heterodyne receiver and signal amplifier, wherein the quadrature detection of registered radiation takes place directly in the active medium prior to the photodetector. When returning to the laser, the reflected wave interferes with an internal wave laser. This is equivalent to the change of the laser resonator losses and of its length. If the gain of the active medium and the resonator losses are close to each other, then a significant change of the weak reflected wave occurs in the output power of the laser, providing the high steepness of the corresponding hardware function. Simultaneously the oscillation frequency varies in the same harmonic dependence on the optical length of the path to the reflector, as for the generation power, but in the quadrature way. The functional features inherent to the autodyne reception enable developing the interferometric devices with performances previously unavailable to the conventional interferometers.

Autodyne interferometers can maintain operability even when virtually any structural part of the plasma installation facing the plasma is used as a reflector.

Using a two-frequency laser in the autodyne interferometer in the mode of cross-exposure of the modes on each other provides high measurement sensitivity, limited by the laser-receiver spontaneous noise [26].

The two-mode Zeeman He-Ne laser is used in the interferometer (λ= 3.39 microns). The two-mode generation mode is realized in an isotropic resonator by superposition of longitudinal magnetic field on the active medium. As a result, the output laser radiation contains two waves with orthogonal circularly polarized modes and the frequency difference *f*, which depends on the magnetic field strength. In the case of double-passing through the quarter-phase plate (Figure 5a) to the reflector and back, radiation modes "share" their polarizations. Thus, the returned light from each mode interacts with the internal wave of the other mode, respectively, with a frequency shift equal to the preset laser intermodal frequency splitting $f=\nu_2-\nu_1$. In such cross-impact events of mode interaction via an external reflector, frequency modulation of intermodal beats and intensities occurs in the laser for both modes $p_{1,2}$ at frequencies *f* and with an amplitude dependent harmonically and quadrature to each other on the path optical length *nL* to the reflector, and linearly on the amplitude reflection coefficient ρ.

$$p_1 \sim \rho P \sin(2knL)\sin(ft) \tag{6}$$

$$\delta f \sim \rho \cos(2knL)\sin(ft) \tag{7}$$

where *P*- the laser generation output, ρ - the reflector amplitude reflection coefficient.

Either the equations for modes strength modulation or their intercarrier frequency modulation contain information as of the effective reflection coefficient ρ and of an optical path length *nL* to the reflector. In (8) the first and the second harmonics are provided for the intermodal beats signal in the approximation of a small modulation index:

$$U_b \sim P\cos(ft) + \frac{1}{2}\rho P \frac{\Delta\Omega^*}{f}\cos[2knL]\cos(2ft) \tag{8}$$

It can be seen that the amplitude of the second harmonic shows linear dependence on ρ and harmonic dependence on *nL*, quadrature relative to the same dependence of the amplitude modulation. Using both dependencies allows determining the value of ρ and optical length *nL* change. To achieve this, radiation from one of the modes is allocated by a quarter-wave plate and the polarizer P_1 (figure 5a). The modulation signal is recorded in the photodetector D_1 and fed through a pre-amplifier to the mixer 3 (figure 5b).

The second photodetector D_2, due to using of a polarizer P_2, detects a beat signal that, after the pre-amping, is divided by narrow-band filters 1 and 2 to the harmonics. A part of the first harmonic is used as a reference signal in a mixer 3. At its output, after carrier *f* is removed, a first informative signal is formed:

$$U_1 \sim \rho P \sin(2knL) \tag{9}$$

Another part of the first harmonic signal is used after the frequency duplicator *4* as a carrier signal in the mixer *5*, where the second harmonic signal is headed. The second informative signal is formed at the output:

$$U_2 \sim \rho P \cos(2knL) \tag{10}$$

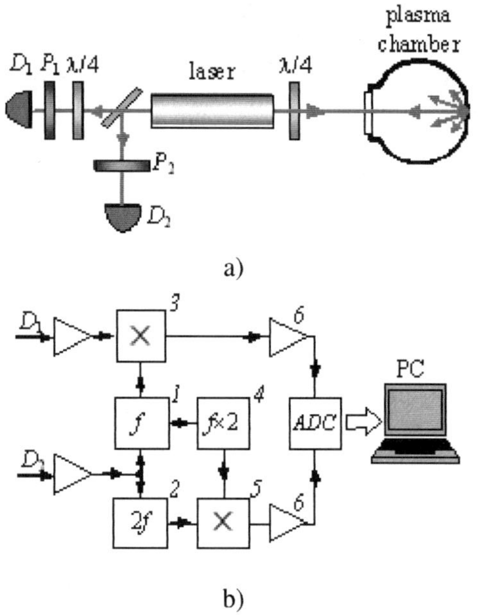

Figure 5. Interferometer optical scheme (a) and informative signal processing functional scheme (b).

Both informative signals pass through low-frequency amplifiers *6* to a high-speed ADC circuit. After digitizing, the data is sent to a computer, where the values ρ and changes in the value of *nL* are calculated.

Thus, according to the nature of the informative signal, the device is a two-channel quadrature interferometer. It serves simultaneously as a lidar, as it allows to record the scattered radiation field and measure the reflection coefficient, and as the interferometer for measuring the change in optical path length.

With the intermodal splitting frequency $f = 20$ MHz the minimal registered signal p, limited by the natural noises level (by the level S/N=1) was $3.5\cdot10^{-7}P\cdot(\mathrm{Hz}^{-1/2})$.

The minimal registered output modulation in the mode complies with the optical length change $(nL)_{\min}=5\cdot10^{-7}\,\lambda$ ($\mathrm{Hz}^{-1/2}$), which corresponds to $N_e\ell=2.5\cdot10^{10}$ ($\mathrm{cm}^{-2}\cdot\mathrm{Hz}^{-1/2}$). The temporal resolution τ of the interferometer is limited by the laser reactive range and equals to 10^{-8} s.

6. Conclusion

Today, it is impossible to imagine plasma physics without a wide range of measurement techniques and parameters control. An important requirement at that point, applied to the measuring techniques, is the absence of contact of the diagnostic instrument with the plasma; otherwise the change of its state may occur. In this regard, non-contact optical methods are mostly preferred in plasma studies. Possessing very distinct dispersive properties, plasma brings phase distortions into a probing electromagnetic wave. Measurements of the phase perturbations with the optical interferometry methods allow obtaining the most precise information about the density of free electrons in plasma.

Quadrature photodetection interferometers provide a wide dynamic range of electron densities measurements while maintaining the homogeneous differential sensitivity. The range of measuring the linear electron density with quadrature interferometers implemented in the pulsed plasma installations is as wide as 10^{10}- 10^{19} cm^2.

The most versatile, possessing the ability to be integrated into a wide class of plasma installations, are the quadrature homodyne interferometers. The sensitivity of the measurements is limited by the photoregistration noises, and the temporal resolution is limited by the speed of photodetection. The use of digital quadrature demodulation algorithms in heterodyne interferometry allows one to get closer to the quantum limit of measurements of small phase shifts, and in autodyne interferometers it is a fundamental property of the active medium, which interrelates the power disturbances and the generation frequency during an injection of a weak radiation into a cavity.

References

[1] Aleksandrov V V et al. 2004 *Plasma Physics Reports* **30** 218
[2] Luhman N C and Peebles W A 1984 *Rev. Sci. Instrum.* **53** 279
[3] Weber B V and Fulghum S F, 1997 *Rev. Sci. Instrum.* **68** 1227
[4] Dooling J C and York T M 1986 *Rev. Sci. Instrum.* **57** 1090
[5] Johnson W B 1967 *IEEE Trans.* **15** 152
[6] Ashby D E T F and Jephcott D F 1963 *Appl. Phys. Letters.* **3** (1) 13
[7] Smith III R S and Dogget W O 1985 *Rev. Sci. Instrum.* **56** 355
[8] Buchenauer C J and Jacobson A R 1977 *Rev. Sci. Instrum,* **48** (7) 769
[9] Greco V, Molesini G and Quercioli F 1995 *Rev. Sci. Instrum,* **66** 3729
[10] Gingston E P *Measurements at centimeter wavelengths* 1960
[11] Schemm J B and Vest C M 1983 *Appl.Opt* **22** 56
[12] Singh H and Sirkis J S 1994 *Appl.Opt* **33** 5016.
[13] Kuznetsov A P, Bashutin O A, Byalkovskii O A, Vovchenko E D, Korotkov K E and Savjolov A S 2008 *Plasma Physics Reports* **34** 193
[14] Kuznetsov A P, Gubskii K L, Protsenko E D, Shapovalov I P and Savjolov A S 2012 *TechnicalPhysics Letters* **38** 1066
[15] Kuznetsov A P, Byalkovskii O A, Gubskii K L, Kozin G I, Protsenko E D, Dodulad E I and Savjolov A S 2014 *Plasma Physics Reports* **40** 290
[16] Kuznetsov A P, Golubev A A, Kozin G I, Mutin T Yu, Savelov A S and Fertman A D 2006 *Instruments and Experimental Techniques* **49** 247
[17] Kuznetsov A P, Byalkovskii O A, Gavrilin R O, Golubev A A, Gubskii K L, Rudskoi I V, Savin S M, Turtikov V I and Khudomyasov A V 2006 *Plasma Physics Reports* **39** 248
[18] Gokay M C and Fusek R L 1981 *Rev. Sci. Instrum.* **52** 1197
[19] Braithwaite G, Gottardi N, Magyar G, O'Rourke J, Ryan J and Veron D 1989 *Rev. Sci. Instrum.* **60** 2825
[20] Kawahata K, Tanaka K and Ito Y 1999 *Rev. Sci. Instrum.* **70** 707
[21] Lamela H and Acedo P 2001 *Rev. Sci. Instrum.* **72** 96
[22] Gurko V F, Zubarev P V, Kvashnin A N and Khil'chenko A D 2003 *Instruments and Experimental Techniques* **46** 619
[23] Kuznetsov A P, Elistratov E A, Koshkin D S, Mikhailyuk A V and Protasov A A 2015 *Instruments and Experimental Techniques* **58** 657
[24] Gerardo J B, Verdeyen J T and Gusinow M A 1965 *J. Appl. Phys.* **36** 2146
[25] 1968 *Plasma diagnostics* ed Lochte-Holtgreven W (Amsterdam: North-Holland Publishing company)
[26] Kozin G I, Kuznetsov A P, Lebedinskii M O, Saveliev A V, Sokolov A P and Savelov A 2003 *Instruments and Experimental Techniques* **46** 188

Interferometry diagnostics of plasma channel of femtosecond filament

P A Chizhov[1,3], V V Bukin[1,2] and S V Garnov[1,2]
[1] A.M. Prokhorov General Physics Institute of the Russian Academy of Sciences, Vavilova Street 38, Moscow, 119991 Russian Federation
[2] National Research Nuclear University MEPhI, Kashirskoe shosse 31, Moscow, 115409, Russian Federation

E-mail: pvch@inbox.ru

Abstract. Results of the electron density dynamics in the plasma channel of a femtosecond filament in air, nitrogen and argon from the moment of ionization up to hundreds of picoseconds after it are presented. Anisotropy of the refractive index was found, which is preceded and accompanied by ionization of the gas at the time of passage of high-intensity femtosecond laser pulse.

1. Introduction

Femtosecond laser plasma is the subject of intense research in recent decades. One of the directions is to study the filamentation process of ultrashort laser radiation [1]. Filamentation phenomenon is the result of a nonlinear optical interaction between the transparent medium and the femtosecond laser pulses of a high power. If the pulse peak power exceeds the critical self-focusing power, the intensity increases with approaching the nonlinear focus. With its increase up to the threshold of photoionization, a laser-produced plasma is formed where defocusing of the light restricts the further growth of intensity in a nonlinear focus. Dynamic balance of Kerr self-focusing and plasma defocusing leads to the stability of the parameters in a long filament.

One of the methods for measuring the electron density and the decay time of the plasma filament during the initial (picosecond) time interval since its birth is the interferometry which uses a weak intensity light beam directed perpendicular to the plasma filament [2].The process of the plasma decay in air goes through several channels of recombination of electrons with various ions including the complex ions [3].

Performing the experiments on the decay of the plasma channel in pure gases allows to reduce the number of possible recombination channels, and thus a simple interpretation of the electron density dependence on time can be expected.

[3] To whom any correspondence should be addressed.

2. Experimental facility

The experimental setup is shown in figure 1. The plasma channel was formed during filamentation of a femtosecond laser pulse (E = 2.5 mJ, t = 40 fs (FWHM), D = 12 mm ($1/e^2$), lens focal length f = 50 cm). This channel was enlightened with a weak probe laser pulse either at the moment of ionization of the medium or after that at a time instant regulated by the optical delay line. The interferograms of the medium with the plasma channel ("signal") and without it ("background") were taken with the CCD camera. The presence of a phase-changing object, the plasma channel, on the way of the probe beam led to a change in the interference pattern. Processing the interferograms with the Fourier filtering method yielded the difference between the "signal" and "background" phase distributions, which corresponds to the contribution of the plasma (Figure 2(a)). Further, assuming cylindrical symmetry of the plasma the refractive index was calculated using the phase shift, and the electron density was evaluated using the Drude model (Figure 2(b)).

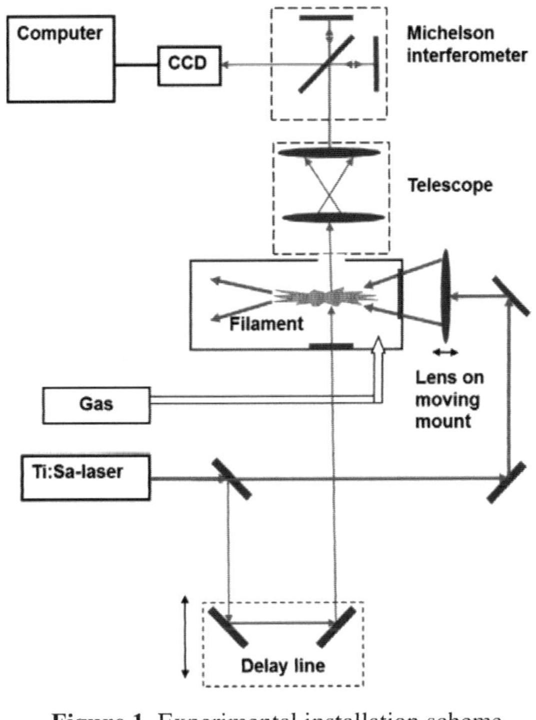

Figure 1. Experimental installation scheme.

3. Results

Thus the time dependence of the electron density during the decay of plasma channel in air and nitrogen was obtained during the time interval of 0-150 ps (Figure 2 (b)). It was found that the decay in argon is much slower than that in nitrogen and air. It was also found that the decay in nitrogen is faster than that in air.

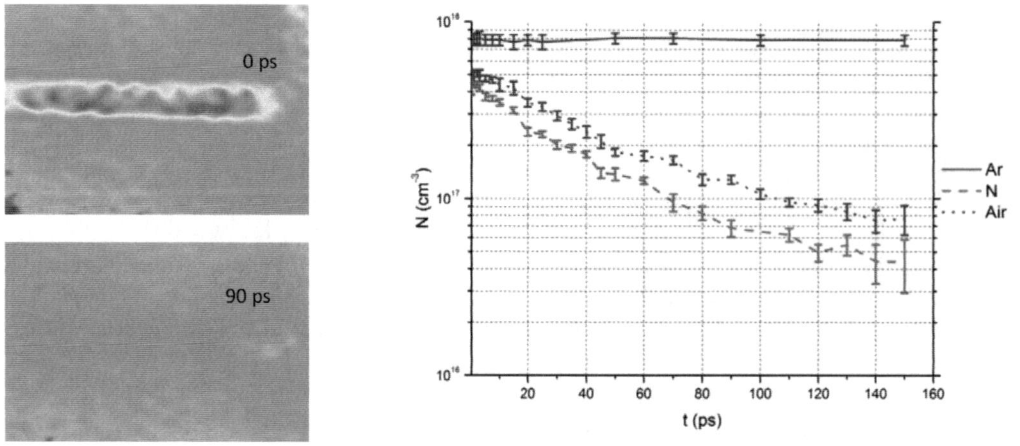

(a) (b)

Figure 2. The images of plasma phase shift at different time instants counted from the moment of ionization (a); the peak electron density in the plasma channel filament as a function of time (b).

Thanks to improvement of interferogram processing method due to averaging over a series of frames, the sensitivity of the method was increased, that allowed to observe the decay in the air up to 1 ns after the moment of ionization (Figure 3).

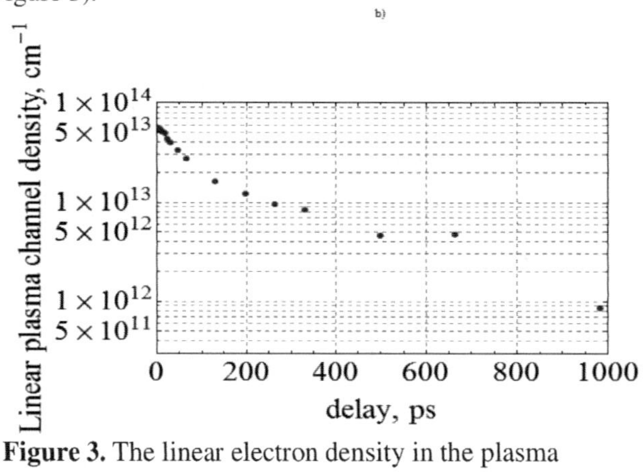

Figure 3. The linear electron density in the plasma channel (air) as a function of time up to 1 ns.

While studying the stage of formation of the plasma channel, anisotropy of the refractive index was discovered, associated with the passage of the high-power laser pulse in air and inert gases (Figure 4). In this case, the phase shift (and the index of refraction) in different regions has both the positive and negative values, that may indicate on the influence of the high-order Kerr effect on the propagation of laser radiation at filamentation.

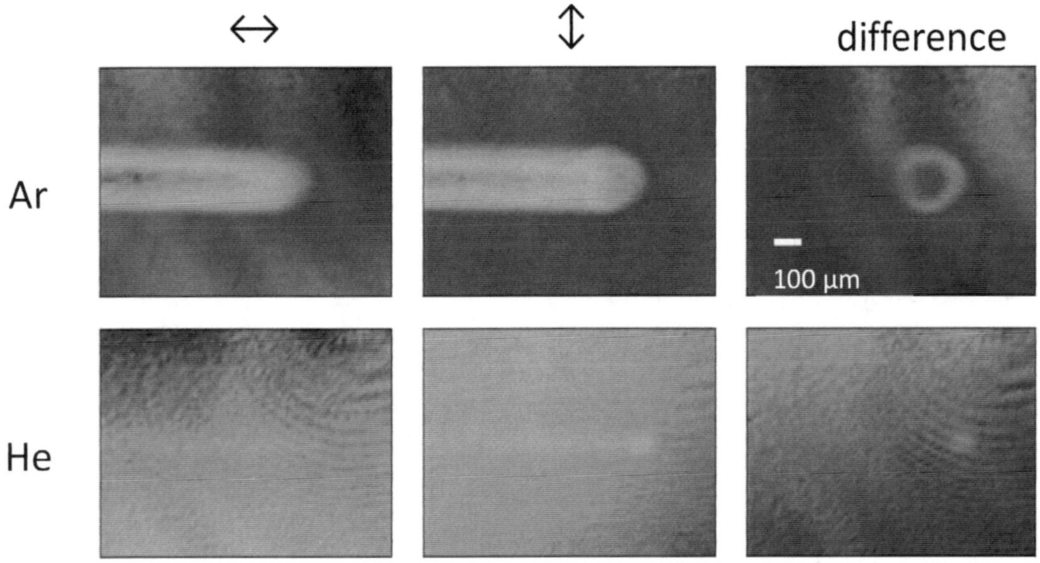

Figure 4. The images of phase shift at the moment of formation of the channel, for various polarizations of the probe beam, which show the anisotropy of the refractive index (↔ - horizontally polarized probe pulse, ↕ - vertically polarized probe pulse, both for vertically polarized ionizing pulse).

Acknowledgments

This work was supported by the Programs "Extreme laser radiation: physics and fundamental applications" and "Fundamental and applied problems of photonics and the physics of new optical materials" of the Presidium of Russian Academy of Sciences, President Grant for Young Scientists and Graduate Students of the Russian Federation (Grant No. SP-2391.2013.2).

References

[1] Couairon A and Mysyrowicz A 2007 *Phys. Reports* **441** 47
[2] Bukin V, Garnov S, Malyutin S and Strelkov V 2007 *Quantum Electron.* **37** 961
[3] Bodrov S, Bukin V, Tsarev M, Murzanev A, Garnov S, Aleksandrov N and Stepanov A 2011 *Opt. Express* **19** 6829

Spectroscopic study of radiation from Low-inductance vacuum spark

D Kirko[1], E Dodulad and A Savjolov
National Research Nuclear University MEPhI (MEPhI), Kashirskoe shosse 31, Moscow, 115409, Russian Federation

E-mail: kirko@plasma.mephi.ru

Abstract. A research of radiation spectra of a low inductance vacuum spark in visible and ultraviolet regions was realized. Time dependences for hydrogen atom lines, and so for iron atom lines and iron ion lines were obtained. Temperature of an external plasma cover was determined.

1. Introduction

The study of low-inductance vacuum spark is performed to simulate characteristic high-temperature plasma processes. X-rays are widely used to analyze behavior of z-pinch plasma[1]. To study the more rarefied peripheral areas of plasma visible and ultraviolet spectra radiation can be used.

2. Experimental setup

We used an electrode system consisted of an anode (3 mm in diameter, made of steel and copper) and a steel cathode (20 mm in diameter) in our experiments. The electrodes are 4-5 mm away from each other. Operating pressure in a chamber is within 10^{-4}-10^{-5} Torr range. The initial voltage is 5-15 kV. The discharge is driven by triggered ignition. The discharge current is within 100-150 kA range with period of 8.5 μs. A 1-2 mm-diameter pinch is formed under the influence of high current. The pinch is surrounded by a more rarefied plasma shell. Furthermore generation of micropinches can be observed in anode's vicinity. The concentration of plasma in pinch area was measured (n_e=10^{17}-10^{19} sm^{-3}) by laser interferometry [2]. X-ray radiation spectrum of pinch and micropinches is comprised of lines of multicharged Fe and Cu ions.

Spectra of vacuum spark in ultraviolet and visible ranges was obtained by Ava Spec 2048 spectrometer (operating range of 200-1000 nm, spectral resolution of 0.3 nm). Monochromator MUM (operating range of 200-800 nm, spectral resolution of 0.2 nm) and photomultiplier PMT-85 (time resolution of 5ns) were used to record temporal dependences.

Steel electrodes were used in the experiments. Radiation was registered from plasma shell of 8-10 mm in diameter. Discharge spectrum in ultraviolet and visible ranges contained a number of intense spectral lines and had a considerable continuous component. Radiation of H_γ 434 nm, H_α 656 nm and H_β 486 nm hydrogen lines was observed (figure 1(a)). These lines are apparently a result of presence of water vapor in residual gas in the facility. Besides that, intense lines of Fe were observed (see figure 1(b)).

[1] To whom any correspondence should be addressed.

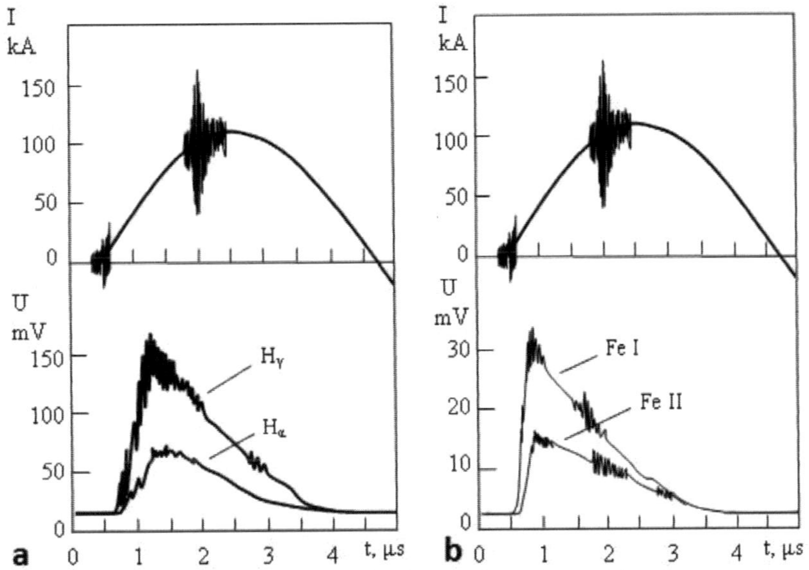

Figure 1. Temporal dependences of spectral hydrogen lines (a) and iron lines (b) intensities.

3. Discussion

The panoramic spectrum of vacuum spark allowed us to obtain general information on basic elemental composition of plasma and integral intensities of the corresponding lines of elements. Temporal dependencies of separate spectral lines were studied to figure out evolution of discharge radiation in time. Figure 1(a) shows typical temporal dependences of H_α 656 nm and H_γ 434 nm. Radiation pulses of these lines have a leading edge of 0.4-0.6 μs at pulse durations of 2.0-2.5 μs. Maximum of intensities of these lines correlate with discharge current maximum. The presence of these hydrogen lines in all the peripheral area of pinch allows diagnostics by relative intensities of spectral lines. It was supposed that pinch shell was in local thermal equilibrium. Calibration of intensities of studied lines was carried out by calibration lamp SIRSH6-100. Using relative intensity method [3] and H lines data, plasma temperature at applied voltage U=14 kV was measured (T=3600±300 K). Radiation of H_α and H_γ lines after decreasing the current to a minimum value may occur due to the presence and radiation of excited hydrogen in the vicinity of the electrodes.

Figure 1(b) shows typical time dependences of spectral lines of atoms and ions of Fe: Fe I 382 nm and Fe II 523 nm. Compared to H lines radiation pulses of Fe lines have steeper leading edge with duration of 0.2-0.3 μs at pulse duration 2.5-3.5 μs. Thus, we can assume that at the initial stage of pinch formation plasma shell is sufficiently transparent for wavelengths of Fe spectral lines. Maximum of Fe lines intensities coincides with the moment of maximal plasma compression which was determined from the discharge current signal.

The spectral measurements of radiation of low-inductance vacuum spark in visible and ultraviolet spectrum ranges carried out in our experiments allow us to determine the composition of plasma shell. Radiation evolution of spectral lines of atoms and single-charge ions was registered. The study of the initial stage of discharge development and plasma shell properties was performed at the moment of pinch formation.

References

[1] Bashutin O A, Savjolov A S and Vovchenko E D 2009 *Plasma Phys. Rep.* **35** 813-817
[2] Savjolov A S, Vovchenko E D and Sivko P A 2000 *Materialy II rossiiskogo seminara "Sovremennye sredstva diagnostiki plasmy"* (M: MEPhI) 46-48 (in Russian)
[3] Huddlstone R and Leonard S 1965 *Plasma diagnostic techniques* (New York: Academic Press) p 515

Diagnostics of Low Temperature Plasma by CO Laser Radiation (5.0-7.5 μm)

Yu M Klimachev[1]
Gas Lasers Laboratory, P.N. Lebedev Physical Institute of the Russian Academy of Sciences, 53 Leninskiy prospect, Moscow, Russia.

E-mail: umk@sci.lebedev.ru

Abstract. A review of the work conducted in Gas Lasers Laboratory of P.N. Lebedev Physical Institute of the Russian Academy of Sciences and devoted to a research of the diagnostic of dynamics of small-signal gain, temperature and population of vibrational levels in the CO-containing gas mixtures excited by pulsed electron-beam sustained discharge is presented. With the help of laser probing techniques the nonlinear Zeeman splitting of the NO lines in a strong magnetic field was also investigated. The studies of gain of the train of nanosecond pulses allowed us to estimate the intensity of the gain saturation of the active medium of the pulsed electron-beam sustained discharge CO laser.

1. Introduction

A significant step forward in the study of kinetic processes occurring in the active medium of electric-discharge CO lasers is the transition from the comparison of the calculated and measured lasing properties of these lasers to a direct comparison of the gain of the active medium, and other parameters of low-temperature plasma. This considerably simplifies the theoretical modeling and reduces the number of parameters measured in the experiment, since it removes the need for modeling the processes of light generation. In addition, the characteristic time of the formation of the population inversion and, consequently, the small-signal gain (SSG), is very important for fast-flow CO lasers with both transverse and longitudinal circulation of the active medium because they have to be considered when developing the design of lasers. Thus, there is a need to study the temporal dynamics of the gain in the active medium of a CO laser under various experimental conditions, in particular, typical for the CO lasers with rapid circulation of the active medium. Also, most important among these conditions (specific input energy and composition of the gas mixture) can be simulated by a pulsed electron-beam sustained discharge (EBSD) CO laser system with stationary cryogenic cooling of the active medium.

In the CO laser the dependence of the SSG of the temperature plays a particularly important role, however, it is very difficult to measure the temperature of the gas directly in the field of electrical discharge. In the paper [1] aimed at measuring the dynamics of the temperature of the active medium

[1] To whom any correspondence should be addressed.

under conditions of electric discharge it was used the fact that the rotational temperature of excited states of the CO molecules became equal to the translational temperature during the time less than 1 ns. The dynamics of the rotational temperature of the gas was evaluated from the dynamics of the SSG for three rotational transitions within a single vibrational band, which was measured using the continuous frequency-selective CO laser as a probe laser. However, the implementation of such a technique with ten percent accuracy of measuring the SSG results in a high error (~200 K) in determining the gas temperature [2]. To increase the accuracy it is necessary, as it follows from [2], to increase the number of probed transitions by several times, i.e. develop a method for the multifrequency probing of active medium of the pulsed CO laser.

2. Multifrequency laser probing of the active medium of the CO laser

In [3, 4] a method of multifrequency laser probing was developed, which was used for the diagnosis of the SSG, temperature and population of vibrational levels in the CO-containing gas mixtures excited by a pulsed EBSD. Optical measuring scheme is shown in figure 1. The dynamics of the SSG at the vibrational-rotational transitions $V+1 \rightarrow V\ P(J)$ of the CO molecule in various mixtures for various specific input energy Q_{in} was experimentally studied [3]. Figure 2 shows the dynamics of small-signal gain in the mixture CO:N_2 = 1:9 for three transitions. It demonstrates that an increase in the vibrational number V the maximum gain G_{max} is reduced, and it is achieved at a later time. The maximum gain G_{max} is reduced threefold for V=31 in comparison with value G_{max} for V=7. It was found that at a fixed value of V the rate of the rise of the SSG is increased with increasing the rotational number.

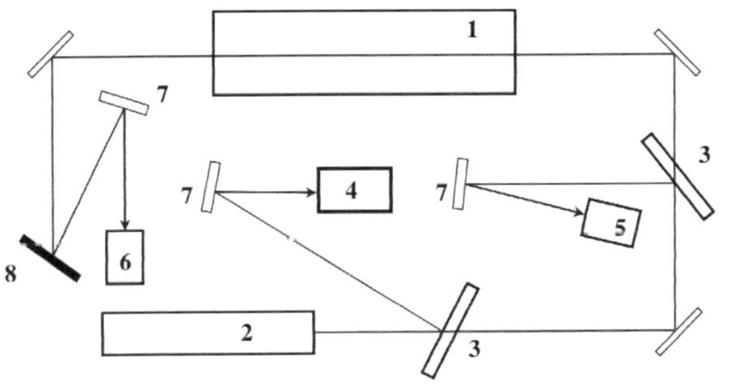

Figure 1. Optical scheme of experiments: (1) laser amplifier; (2) probe laser; (3) beam splitter; (4) IR spectrometer; (5, 6) photodetectors; (7) spherical mirror; (8) diffraction grating.

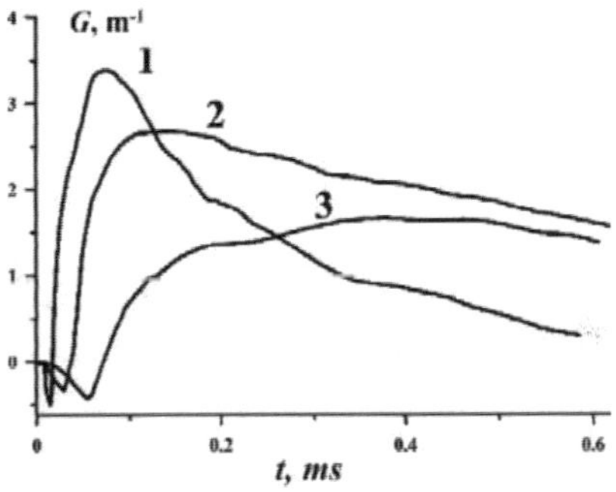

Figure 2. The SSG dynamics in the CO:N_2=1:9 mixture at 8→7 P(10) (**1**), 14→13 P(10) (**2**) and 21→20 P(11) (**3**) transitions for Q_{in} =800 J L^{-1} $Amagat^{-1}$.

It was shown that for the oxygen-containing gas mixtures (CO:O_2 = 1:19) the value G_{max} at lower vibrational transitions (for V <13) can be significantly greater than G_{max} for the mixture in which the nitrogen is used instead of oxygen (CO:N_2 = 1:19). It has been found that the efficiency of the CO laser operating on a mixture with higher oxygen content is increased up to 47% in comparison with the efficiency of the CO laser operating on the nitrogen-containing gas mixture (where it equals to 30%).

It has been demonstrated that the method can reduce the error in the determination of the gas temperature up to 3% [4]. It was found that in the mixture of CO:O_2 the population of lower vibrational levels of the CO molecule can be several times higher than the population of respective levels in the laser mixtures CO:He and CO:N_2. However, the temperature rise in the oxygen-containing mixtures is significantly higher than that in helium and nitrogen mixtures. Figure 3a shows the time dependence of the absorption for three transitions of the CO molecule, on the basis of which the temperature dynamics has been determined (figure 3b).

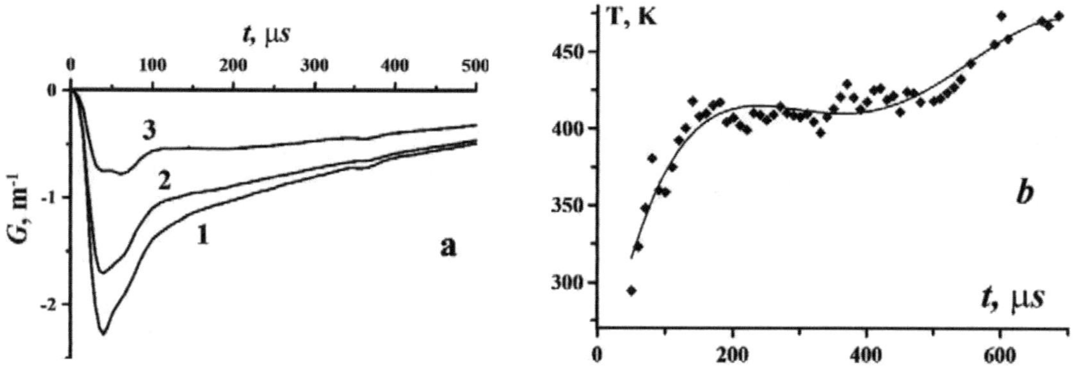

Figure 3. Absorption coefficient of the probe radiation (**a**) at the transitions in the 9→8 P band for J=9 (**1**), 11 (**2**), and 15 (**3**), and the temperature dynamics (**b**) of the CO:O_2=1:19 gas mixture for pressure 30 Torr, initial temperature 290 K, and Q_{in} = 250 J L^{-1} $Amagat^{-1}$.

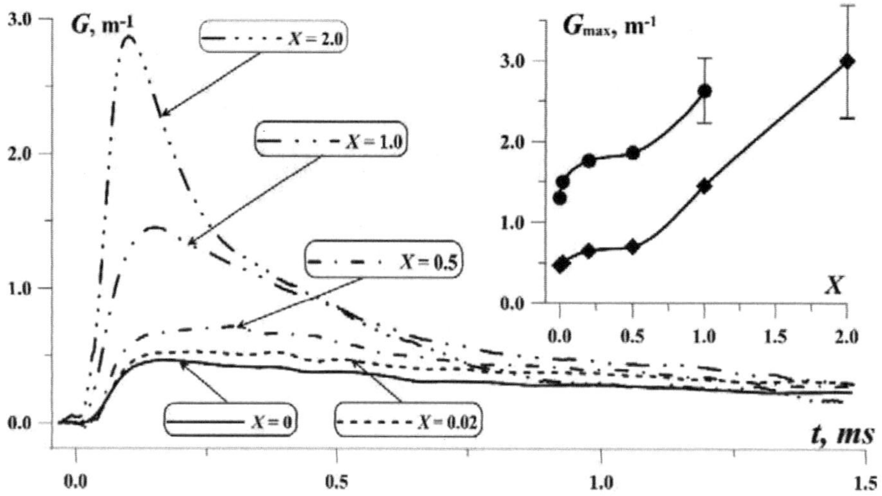

Figure 4. The SSG dynamics in the CO:He:O_2=1:4:X mixture at the 10→9 P(15) transition for five values of X (Q_{in} = 130 J L^{-1} $Amagat^{-1}$). The inset shows the dependence of G_{max} on X for Q_{in} = 240 (•) and 130 J L^{-1} $Amagat^{-1}$ (♦). The measurement error is within the marker size, except the shown error bars.

Further research showed [5] that the addition of molecular oxygen to the working mixture of a pulsed EBSD CO laser amplifier multiplies the SSG of the active medium in the vibrational transitions of the CO molecule for *V* from 6 to 13 (figure 4), which is associated with an increase of the population of the vibrational levels. However, the lifetime of the gain decreases for all the probed transitions.

Figure 5 shows the vibrational energies of transitions of O_2, CO and N_2 molecules. This figure demonstrates the essential role of kinetic processes of intermolecular vibrational-vibrational (VV') exchange for the excitation of the CO molecules. It was found that the addition of oxygen increases the efficiency of the CO laser at the fundamental transitions, and the efficiency reaches its maximum value at lower values of specific input energy. It has been shown that the pulsed EBSD CO laser can operate on the air working mixture both at the fundamental transitions and the first overtone vibrational transitions V+2→V.

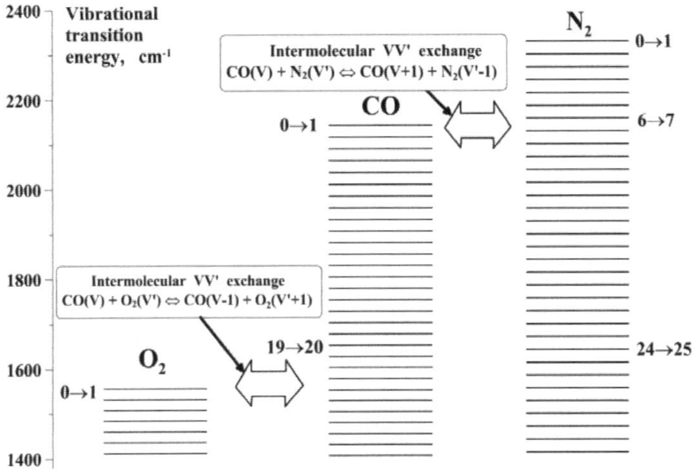

Figure 5. Vibrational transition energies of O_2, CO, and N_2 molecules.

In the paper [6], we researched the influence of small additions of O_2 on the dynamics of small-signal gain of the active medium of a pulsed EBSD CO laser at cryogenic temperature. Using a numerical model of the laser active medium has enabled us to find out how the VV'-exchange between vibrationally excited molecules at high levels of CO and oxygen molecules affects the dynamics of small-signal gain. Methodology based on a comparison of the measured and calculated dynamics of small-signal gain for a number of transitions (figure 6) enabled us to determine the parameters of the analytical approximation of the rate constants of the VV'-exchange for the pair of molecules CO-O_2 at cryogenic temperature of the active medium.

3. The research of the Zeeman splitting of the IR absorption lines of the NO molecule

In the paper [7], the dependence of the Zeeman splitting of the vibrational-rotational lines of the NO molecules on the value of a strong magnetic field with magnetic induction up to 6 Tesla was theoretically and experimentally studied. Measurement of absorption in a pulsed magnetic field was also carried out using a CW frequency-tunable CO laser operated in a single line selective mode. To analyze the dynamics of absorption, a numerical model of the nonlinear Zeeman splitting of the NO lines has been developed. The NO molecules absorption dynamics in a pulsed magnetic field was described by the calculation model in a cubic approximation of the value of magnetic induction. A non-linearity of the Zeeman splitting lines of the Q- and P-branches of the vibrational-rotational transitions of the NO molecules in the $^2\Pi_{3/2}$ state was found. Particularly, for the lines Q(2.5), P(5.5) and P(6.5) the non-linearity of the Zeeman components of the lines was identified for the magnetic induction of ~4 T. Comparison of the calculated and measured absorption dynamics of the probe radiation showed that the calculation model reproduces well the position of the measured absorption peaks depending on the induction of a pulsed magnetic field (figure 7), that allowed, thus, to identify the Zeeman components of the absorption lines of the NO molecules.

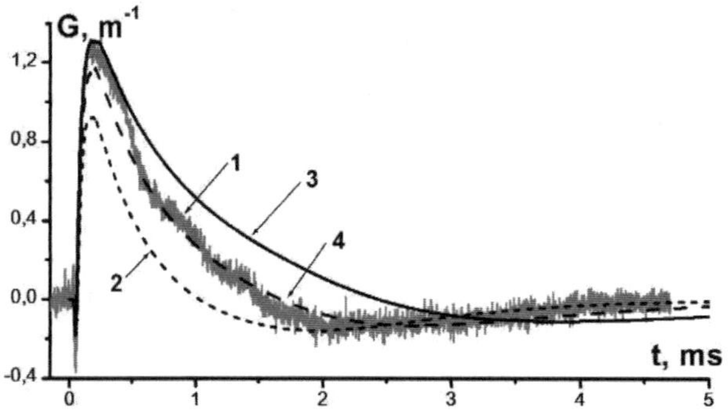

Figure 6. The SSG dynamics at the 18→17 P (15) transition for Q_{in} = 240 J L^{-1} $Amagat^{-1}$ in the gas mixture $CO:He:O_2$=1:4:0.05. Experimental (1) and calculated values (2-4) are shown for various values of the VV' exchange rate constant.

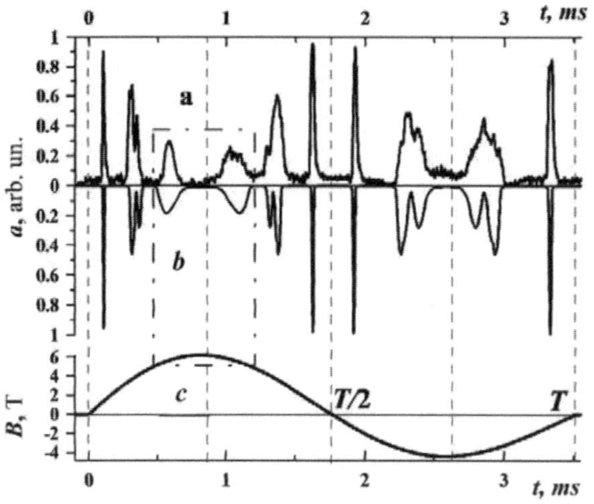

Figure 7. Experimental (a) and calculated (b) absorption dynamics $a(t)$ for the NO molecules for the CO laser line 9→8 P(15) in a damped oscillating magnetic field (c). Time instant corresponding to the value of B as high as 5 T is marked by the dash-dot lines.

4. The research of the CO laser nanosecond pulses amplification

The works [8, 9] were dedicated to a study of the active mode-locking of the EBSD CO laser, creating a master oscillator and laser amplifier system emitting a train of nanosecond pulses, as well as to a research of nanosecond pulse train amplification in the active medium of the CO laser amplifier. The duration of the spikes was ~10 ns, with the minimum value of the duration, achieved in our experiments, equal to ~5 ns. We got the dependence of the output laser energy of the amplifier from the input laser energy delivered to the amplifier for various values of the specific input energy delivered to the EBSD and of the relative density of the working gas mixture, which had a clearly non-linear character. This fact indicates the saturation of amplification of the active medium.

Maximum peak power P_{peak} of the radiation amplified in one pass for the CO : He = 1 : 4 mixture was 380 kW (figure 8) for the 3.2 J total energy of a train of nanosecond pulses at the output of the laser amplifier in the nonselective mode and 100 kW for, respectively, 1.1 J in the frequency-selective mode. At the same time the efficiency of the "master oscillator-laser amplifier" system compared to the pure laser mode of forming a train of nanosecond pulses increased by a factor of 2 in the selective mode, reaching 1.6%, and, respectively, by a factor of 1.5 for the non-selective mode, amounting to 5.3% [9]. To estimate the saturation parameter, the rotational relaxation time of the CO molecules was calculated for our conditions, and amounted to 1.7 ns that is significantly shorter than the duration of the radiation spikes. By comparing the experimental and calculated dynamics of the radiation spikes at the output of the laser amplifier the value of the saturation intensity of nanosecond pulses amplification was found in the active medium, amounting I_S = 14±4 kW/cm^2 for the vibrational-rotational transition of the CO molecule 9-8 P (11) at Q_{in} = 210 J/(L Amagat).

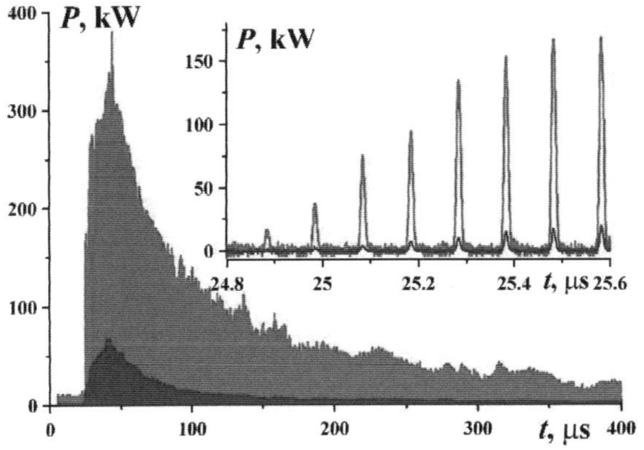

Figure 8. The power of a train of nanosecond pulses at the input (blue) and the output (red) of laser amplifier for nonselective mode (~ 10 spectral lines in the range from 5.1 to 5.4 microns). The inset shows the beginning of the train of nanosecond pulses, illustrating the shape and repetition frequency of the spikes.

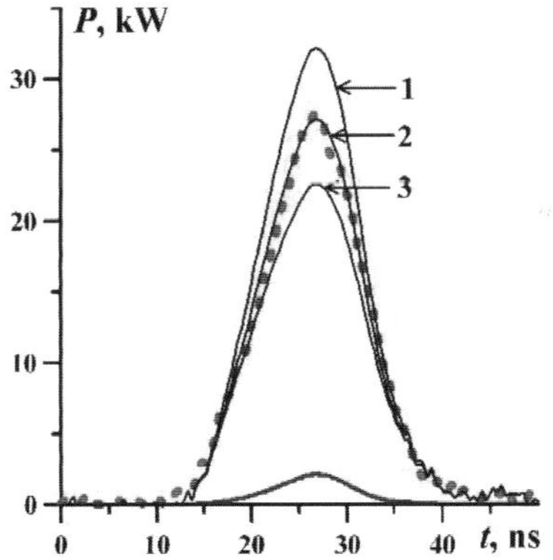

Figure 9. Experimental data for dynamics of the radiation spike at the input (blue) and output (red) of the amplifier at t = 37 μs, and the calculation results (black curves) for the values of gain saturation intensity I_S = 20 kW cm^{-2} (1), I_S = 15 kW cm^{-2} (2), and I_S = 10 kW cm^{-2} (3).

5. Conclusion

Developed and applied in the Gas Lasers Laboratory of P.N. Lebedev Physical Institute of the Russian Academy of Sciences, a method of multifrequency laser probing enabled us to hold a series of studies on the diagnosis of the dynamics of the small-signal gain, temperature and population of vibrational levels in the CO-containing gas mixtures excited by a pulsed electron-beam sustained discharge. With its help we also studied the non-linear Zeeman splitting of the NO lines in a strong magnetic field. The researches of amplification of a train of nanosecond pulses enabled us to estimate the intensity of the gain saturation in the active medium of a pulsed electron-beam sustained discharge CO laser.

This work was partially supported by Russian Foundation for Basic Research (grant 13-02-01135).

References

[1] Basov N G, Kazakevich V S, Kovsh I B and Mikryukov A N 1983 *Soviet Journal of Quantum Electronics* **13** 667
[2] Vyazovetskii N M, Didyukov A I, Kirko V Yu, Kulagin Yu A and Shelepin L A 1990 J. *Appl. Spectr.* **52** 444
[3] Vetoshkin S V, Ionin A A, Klimachev Yu M, Kozlov A Yu, Kotkov A A, Rulev O A, Seleznev L V and Sinitsyn D V 2007 *Quantum Electronics* **37** 111
[4] Ionin A A, Klimachev Yu M, Kozlov A Yu, Kotkov A A, Rulev O A, Seleznev L V and Sinitsyn D V 2007 *Quantum Electronics* **37** 231
[5] Ionin A A, Klimachev Yu M, Kozlov A Yu, Kotkov A A and Seleznev L V 2008 *Quantum Electronics* **38** 115
[6] Ionin A A, Klimachev Yu M, Kozlov A Yu, Kotkov A A, Kurnosov A K, Napartovich A P and Shnyrev S L 2008 *Quantum Electronics* **38** 833
[7] Ionin A A, Klimachev Yu M, Kozlov A Yu and Kotkov A A 2011 *J. Phys. B: At. Mol. Opt. Phys.* **44** 025403
[8] Ionin A, Kinyaevskiy I, Klimachev Yu, Kotkov A and Kozlov A 2012 *SPIE Newsroom*, doi: 10.1117/2.1201112.004016
[9] Ionin A A, Kinyaevskiy I O, Klimachev Yu M, Kozlov A Yu and Kotkov A A 2012 *Optics Communications* **285** 270

CCD camera-based diagnostics of optically dense pulsed plasma with account of self-absorption

I S Nikonchuk[1, 2] **and A N Chumakov**[1]
[1] B.I. Stepanov Institute of Physics, National Academy of Sciences of Belarus, 68 Prospekt Nezavisimosti, Minsk BY-220072, Republic of Belarus

E-mail: i.nikanchuk@dragon.bas-net.by

Abstract. We present a system developed for the optically dense pulsed plasma diagnostics based on a digital video camera with a CCD matrix, that provides determination of spectral brightness and optical density with account of self-absorption.

1. Introduction

Real sources of a pulsed plasma (lasers, discharges, etc.) are often not optically thin, so the diagnostics of such plasmas should be performed taking into account the reabsorption of the radiation in the investigated source [1]. Plasma self-raying by its own radiation [2, 3], or by the radiation from an external source [4] can be used to determine the optical density and spectral absorption coefficients of the investigated plasma. The special scheme of self-raying, that provides spectral transmission factors with a resolution in any dimension of the source in a single experiment in combination with the spectrograph allows determining the temperature and transmission factors of plasma [3]. We used this scheme to create a prototype of a diagnostic system for measuring the spectral brightness and optical density of a pulsed plasma with a CCD digital video camera. This allows improving the accuracy of measurements, minimizing the size of the experimental setup, as well as speeding up and simplifying the data recording and processing.

2. Experimental details and analysis

The prototype of the diagnostic system schematically shown in figure 1 consists of a plasma self-raying device and optical radiation registration system based on the SONY ICX415AL CCD image sensor. A special mask *2* is a lattice with uniformly positioned slits, made of a material that absorbs optical radiation, placed on the surface of the concave spherical mirror *1* of the self-raying device. The plasma source *5* generates the plasma torch *4*. The objective lens *3* projects an image of the torch to the mirror *1*. The radiation reflected from the mirror allows the lens to form the real image of the plasma source, that coincides with the original image. The objective lens *7* projects the image of the plasma torch through the system of optical filters *6, 8* on the CCD *9*. The "CCDImage" software is used for setting the registration parameters, saving and exporting the results in various formats for further data processing.

[2] To whom any correspondence should be addressed.

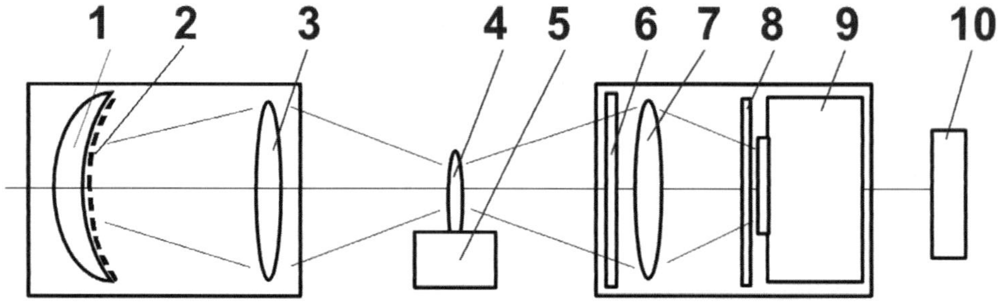

Figure 1. Optical self-ray diagnostic system. 1 – concave spherical mirror; 2 – mask; 3 – objective lens; 4 – plasma torch; 5 – plasma source; 6-9 – digital video camera (6 – neutral filters set; 7 – objective lens; 8 – narrow-band filter; 9 – optical registration system based on the CCD Image Sensor SONY ICX415AL); 10 – computer.

The intensity distribution of the plasma torch registered by the diagnostic system has the following form: the amplitude of the signal from the plasma object is registered at the position of the absorbing strips of the mask, and the amplitude of the signal from the self-rayed plasma can be estimated between these strips (figure 2).

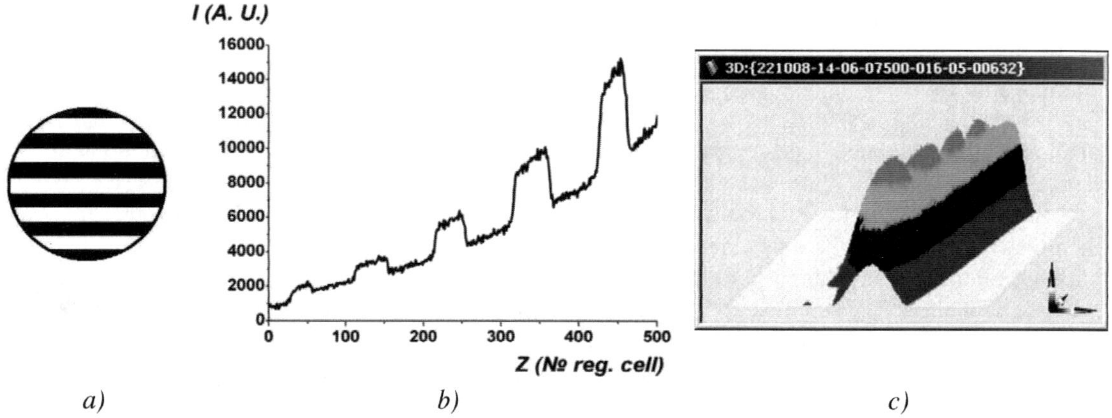

Figure 2. The configuration of the mask (a). The intensity distribution (2D (b) and 3D (c)) of the self-rayed plasma radiation, using the hydrocarbon burner flame as an example.

Digital camera based on the SONY ICX415AL CCD array was equipped with narrow-band and neutral filters and calibrated according to the spectral radiance of the SI10-300u photometric lamp metrologically certificated in the 0.3 - 2.5 μm range [5]. The SIP-30 stabilized DC unit was used as a lamp power supply. The current was measured with the M-104 ammeter.

The spectral brightness of plasma can be found using the spectral radiance of the reference radiation source and registration conditions (exposure and the plasma glow time, aperture, shooting distance and transmittances of filters) [5]:

$$B_1 = B_2 \frac{\Omega_2}{\Omega_1} \frac{S_2}{S_1} \frac{\tau_2}{\tau_1} \frac{I_1}{I_2} \qquad (1)$$

where B is the spectral brightness (1 – plasma, 2 – reference source), S is the cross-section of the radiation source, Ω is the solid angle, τ is the transmission coefficient of the set of filters and I is the signal amplitude.

Assuming that the solid angle and the cross-section of the radiation source are given by

$$\Omega = \frac{\pi}{4}\frac{f^2}{D^2 \cdot L^2}$$

and

$$S = K^2 \cdot d_{pixel}^2$$

formula (1) (when using the same lens) becomes

$$B_1 = B_2 \frac{D_1^2}{D_2^2}\frac{L_1^2}{L_2^2}\frac{\tau_2}{\tau_1}\frac{\Delta t_2}{\Delta t_1}\frac{K_2^2}{K_1^2}\frac{I_1}{I_2} \tag{2}$$

where B is the spectral brightness (1 – plasma, 2 – reference source), D is the f-number (the ratio of the lens's focal length to the diameter of the entrance pupil) , L is the distance from the lens to the object, τ is the transmission coefficient of the set of filters, Δt is the exposure time, K is the magnification of the optical registration system, and I is the signal amplitude.

Figure 3 shows examples of optical characteristics determination of the plasma torch formed by the free-running Nd:YAG laser (λ = 1064 nm, τ = 200 μs, pulse energy up to 1 J, beam diameter of 8 mm) irradiation of a structural steel target.

The transmission coefficient of the optical system is given by:

$$B1'/B1 = 1 + \rho\lambda\tau \tag{3}$$

where B_1 is the brightness of the radiation of the plasma torch, B_1' *is* the brightness of the plasma radiation, enhanced by the reflected beam, ρ_λ is the transmittance of the optical scheme based on the calibration, τ is the transmittance of the plasma (*B1* and *B1'* are calculated by the formula (2)).

Using (3) we can derive the transmittance, and then the optical density and the absorption coefficients of the plasma. The obtained values of spectral radiance of the plasma source can be used not only to calculate its radiance temperature, but also to assess the actual temperature on the basis of the obtained absorption coefficients of the plasma and the use of the Kirchhoff's law.

Test experiments with the plasma torch formed by the Nd:YAG laser irradiation of targets made of different materials, showed that equipping a digital camera calibrated to the absolute value of the radiance in narrow spectral intervals with the plasma self-raying device significantly enhances the diagnostic capabilities of the investigated plasma, taking into account its reabsorption in the plasma optical density within 0.1 to 1.3 range.

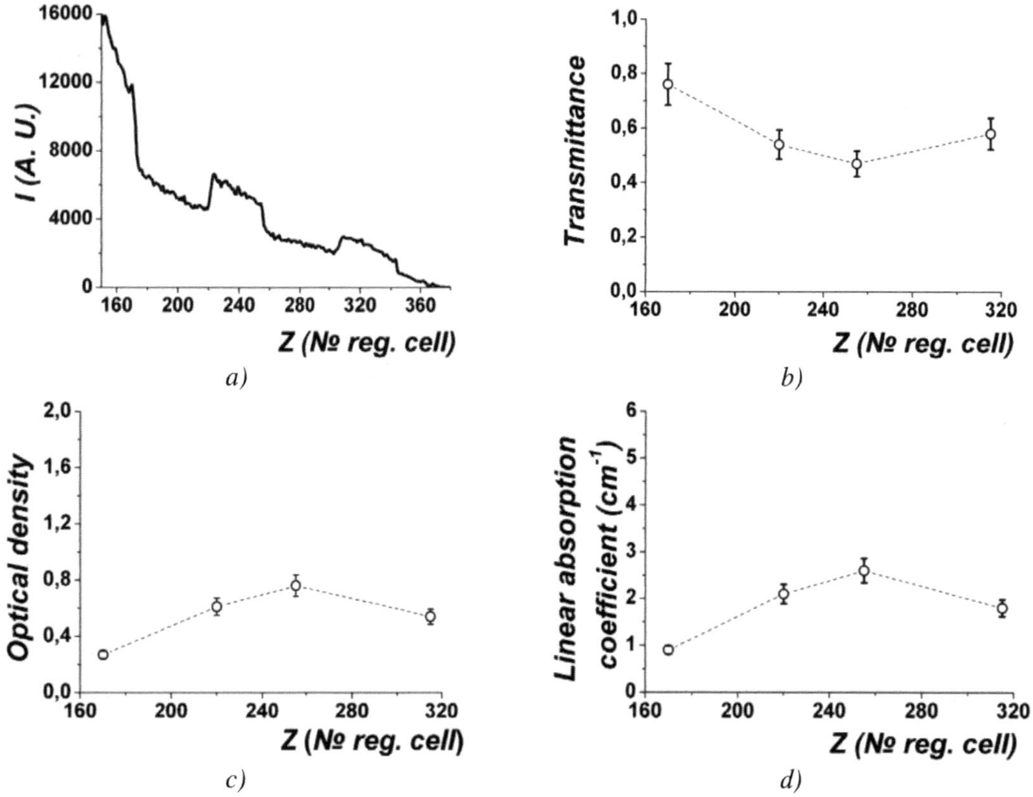

Figure 3. Distribution of the radiation intensity of the plasma formed by the impact of the laser radiation on the steel and illuminated by its own radiation (a); optical characteristics of the plasma torch (b-d).

3. Conclusion

The prototype of the diagnostic system based on the CCD digital video camera and the plasma self-raying device allows increasing the accuracy of the spectral radiance and the optical density measurements of a pulsed plasma, minimizing the size of the experimental setup, as well as speeding up and simplifying the data recording and processing.

References

[1] Lochte-Holtgreven W 1968 *Plasma diagnostics* (Amsterdam: North-Holland) 932
[2] Podmoshenskij I and Shelemina V 1959 *Optika i Spectroscopija* **6** 813 (in Russian)
[3] Goncharov V K et al 1973 *Sov. J. Quantum Electron.* **3** 29
[4] Fedorovich O A 2008 *Jaderna fizika ta energetyka* **2** 86 (in Russian)
[5] Nikonchuk I and Chumakov N 2012 *Proc. of IX Belorussian-Serbian symposium "Physics and diagnostics of laboratory and astrophysical plasmas* (Minsk) 114

Optogalvanic control of instabilities in dusty plasma

M Y Pustilnik[1,5], A V Ivlev[2], N Sadeghi[3], R Heidemann[2], S Mitich[4], H M Thomas[1] and G E Morfill[2]
[1] German Aerocosmic Centre, Complex Plasma Research Department, Wesling, Germany
[2] Max Planck Institute for Extraterrestrial Physics, Garching, Germany
[3] Interdisciplinary Physics Laboratory, Joseph Fourier University Grenoble, France
[4] Justus Liebig University Giessen, Germany

E-mail: Mikhail.Pustylnik@dlr.de

Abstract. This paper presents experimental results on the control of periodic contraction of the void border and proposes a hypothesis on its mechanism, based on the distinctive features of a dusty plasma - its heterogeneity and the presence of the volumetric sink in it.

Instabilities in low-temperature plasmas with negative ions are known for decades [1, 2]. They appear in the form of low-frequency (1-100 kHz) spontaneous discharge fluctuations. Usually, they are global - the discharge glow demonstrates in-phase changes throughout the discharge. These instabilities can be described on the assumption of a homogeneous distribution of the parameters over the volume of the discharge.

Dusty plasma can also be considered as a plasma with very heavy negative ions. However, the macroscopic nature of dust particles adds some features. First, the charge of the dust particles is determined by balance of the flows of electrons and ions onto their surface. The surface of the dust particles, therefore, is a volumetric plasma sink. Secondly, for the dust particles the ion drag force becomes significant. Its presence leads to the formation of so-called voids – cavities free from the dust particles [3]. Such plasma cannot be considered homogeneous.

In the first dusty plasma experiments in microgravity, under certain conditions, spontaneous periodic contraction of the void border was observed. This phenomenon is called "heartbeat". The mechanism of appearance of this instability is still not clear. This paper presents experimental results on the control of this instability and proposes a hypothesis on its mechanism, based on the distinctive features of a dusty plasma - its heterogeneity and the presence of the volumetric sink in it [4].

Experiments were performed in laboratory conditions in the discharge chamber of the earth model of the orbit experiment, PK-3-plus [5]. Monodisperse microspheres of melamine-formaldehyde resin with a diameter of 1.95 micron were injected into the volume of the capacitive HF discharge (argon, pressure of 10-50Pa). Gravity affecting them was compensated by the thermophoresic force: a temperature difference of 19K was maintained between the electrodes. To control the instability, a tunable semiconductor laser utilizing argon transition with the wavelength of 772.38 nm was used. The output power of the laser beam was varied by attenuation through the filters. The maximum

[5] To whom any correspondence should be addressed

power was 8 mW. The beam was modulated with a mechanical chopper. The experimental scheme is shown in figure 1.

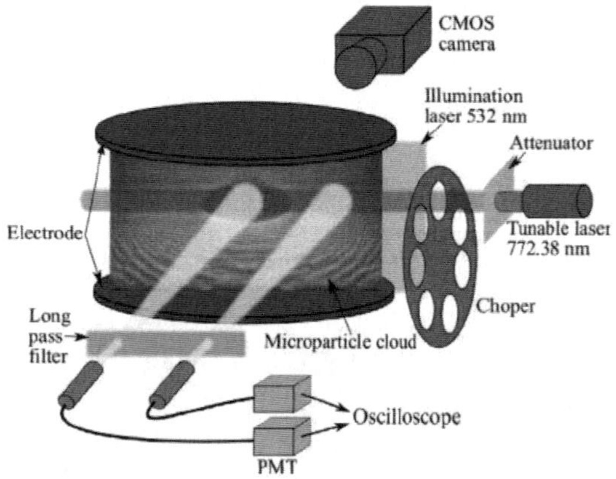

Figure 1. Experimental setup for studying the “heartbeat” in dusty plasma.

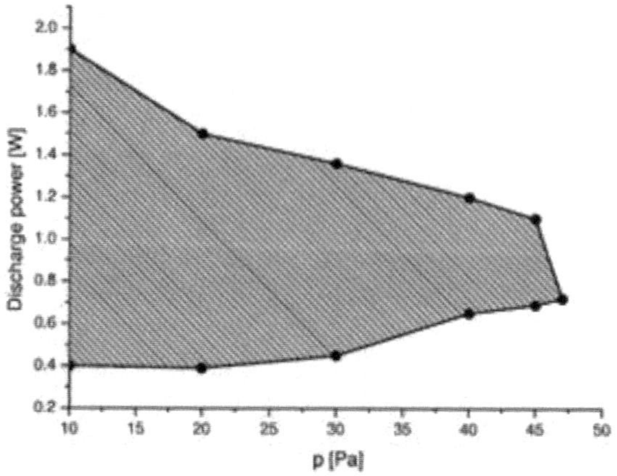

Figure 2. Typical diagram of stability of the dust particle system in a capacitive HF discharge with respect to the “heartbeat”.

A typical stability chart for the system of dust particles with a void in a capacitive HF discharge is shown in Figure 2. The experiments on controlling the "heartbeat" were performed at the pressure of 20 Pa near the lower (in power) border of the instability region. The discharge was brought to a stable state by lowering the discharge power by 0.1-0.2 W below the threshold.

Figure 3a shows the dependency of the "heartbeat" rate on the frequency of the laser beam modulation at the constant power of 3.5 MW. It indicates that excitation of the "heartbeats" has a resonant nature. Figure 3b shows the dependency of the "heartbeat" rate on the laser output power at the constant modulation frequency of 55 Hz. Fourier analysis of the intensity of plasma emission from the void allows one to assume that laser excitation of the "heartbeats" has the character of a parametric instability.

Figure 4 shows the dependency of the "heartbeat" rate on the fraction of the volume of the void illuminated by the laser beam. One can see that it is the impact on the area of the void that is responsible for laser excitation of the "heartbeats.

Therefore, the mechanism of occurrence of the "heartbeat" must include some critical phenomenon inside the void. The hypothesis presented in this paper is that this critical phenomenon is the appearance of a near-electrode sheath-like structure on the border of the void.

In fact, the inner surface of the void is a porous wall. If this wall is sufficiently loose, the plasma losses occur mainly on the inner surfaces of the discharge chamber. The void border does not appear to be a special point. However, if the density of the dust component is being gradually increased, it will eventually reach the level at which the plasma loss at the void border is so great that the drift velocity of the ions reaches the Bohm speed. Then, the quasi-neutrality near the void border will inevitably be broken and the sheath will appear. The electric field at the void border will increase dramatically, while the ion drag force will not change much (the ion flux changes continuously). This will lead to the observed collapse of the void.

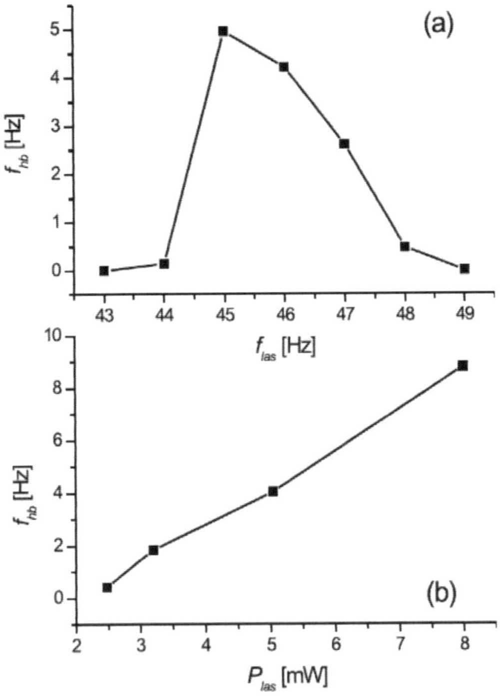

Figure 3. Dependency of the "heartbeat" rate (a) on the frequency of modulation of the laser beam with the permanent emission power of 3.5 MW and (b) on the emission power at the constant modulation frequency of 55 Hz.

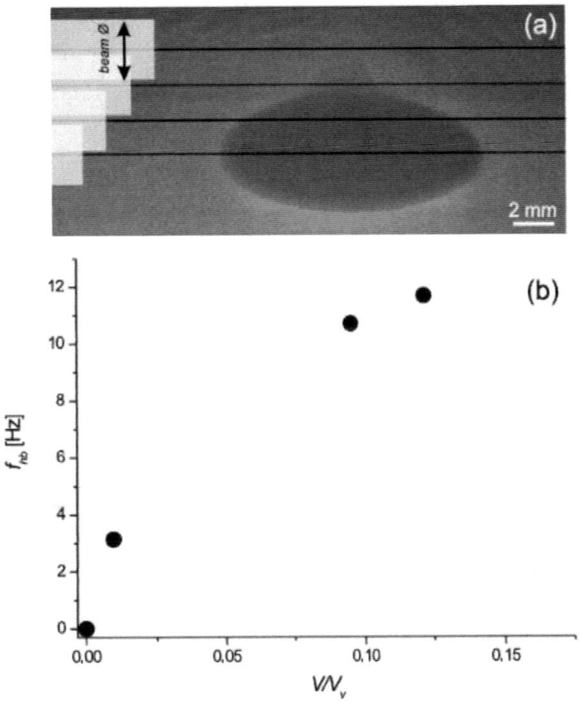

Figure 4. Dependency of the "heartbeat" rate on the laser beam coverage of the void area.

Estimates of the initial velocity of the dust particles during the collapse, the plasma parameters and the dynamics of the radiation intensity made based on this hypothesis confirm its plausibility, although no direct formation of the sheath at the void border by excitation of the "heartbeat" has been observed yet.

References

[1] Alexandrov N L, Napartovich A P 1993 *Uspekhi Phys. Nauk* **163** 1 (in Russian)
[2] Chabert P, Lichtenberg A J, Lieberman M A, Marakhtanov A M 2001 *Plasma Sources Sci. Technol.* **10** 478
[3] Morfill G E, Thomas H M, Konopka U, Rothermel H, Zuzic M, Ivlev A, Goree J 1999 *Phys. Rev. Lett.* **83** 1598
[4] Pustylnik M Y, Ivlev A V, Sadeghi N, Heidemann R, Mitic S, Thomas H M, Morfill G E 2012 *Phys. Plasmas* **19** 103701
[5] Thomas H M *et al* 2008 *New J. Phys.* **10** 033036

Diagnostic suite for study of corpuscular flow dynamics in ion-optical system of neutron tube

A N Dolgov[1], V G Markov[1], I A Kanshin[1,3], D E Prokhorovich[1], A G Sadilkin[1], I V Vizgalov[2], V I Rashchikov[2], N V Mamedov[2] and D V Kolodko[2]
[1] Federal State Unitary Enterprise «All-Russia Research Institute of Automatics», Sushchevskaya street 22, Moscow, 127055, Russian Federation
[2] National Research Nuclear University MEPhI, Kashirskoe shosse 31, Moscow, 115409, Russian Federation

E-mail: vniia4@vniia.ru, ilia.kanshin2011@yandex.ru

Abstract: In this work we developed a combined approach based on the experimental investigation of corpuscular flow at the outlet of an ion source and the computer simulations of beam transport inside ion-optical system that enabled us to recover the parameters of neutron tube operation.

A diagnostic suite was used to study the parameters of corpuscular flow emitted by a Penning-type ion source. The suite includes a single Langmuir probe, a three – electrode – grid electrostatic probe, a Faraday cup assembly, an emittance meter, and a system of corpuscular beam trace imaging in the accelerating gap. The use of those devices enabled us to register the following parameters of corpuscular flows at the outlet of the ion source: electron temperature and ion density in the plasma escaping from the ion plasma source, the distribution of ions in the kinetic energy, total current of emitted ions and the distribution of the current density across the cross-section of the ion flow; emittance diagrams, and the trajectory of corpuscular beam in the accelerating gap.

The experiment for evaluation of electron temperature involved measuring the current flowing through a single Langmuir probe at different potential biasing with respect to the grounded walls of vacuum chamber, i.e. taking the current voltage (*I-V*) characteristics of the probe. Measuring the steep part of the *I-V* curve made it possible to evaluate, from the tilt angle, the electron temperature $T_e = (5.0 \pm 1.2)$ eV [1]. The probe current in the "ion saturation" regime enabled us to make an order-of-magnitude estimate of the ion density: $n_i \approx 10^8$ cm^{-3}.

The study of energy spectrum of corpuscular flow was conducted using the retarding field method with the help of a three – electrode probe (figure 1) located at the outlet of the ion source instead of the Langmuir probe. The probe was comprised of two flat tandem grids and a collector (Faraday cup). Registration of retardation of the ion component of the corpuscular flow was performed by varying the collector's potential bias with respect to the first grid which was grounded. The second grid had a negative potential (-60 V) with respect to the grounded first grid (for cutting off the electron plasma component).

[3] To whom any correspondence should be addressed.

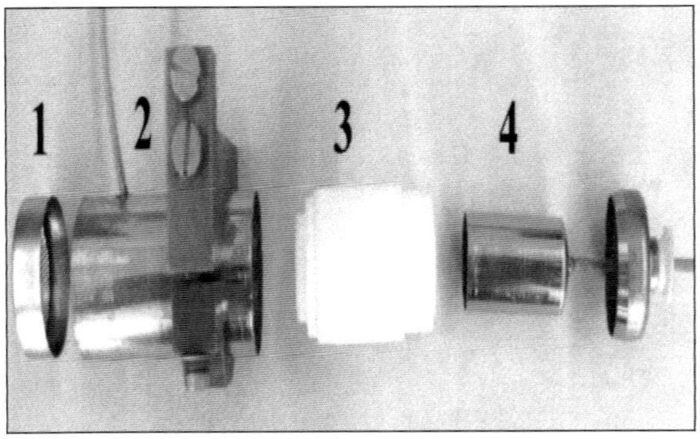

Figure 1. The view of the disassembled three-electrode probe.

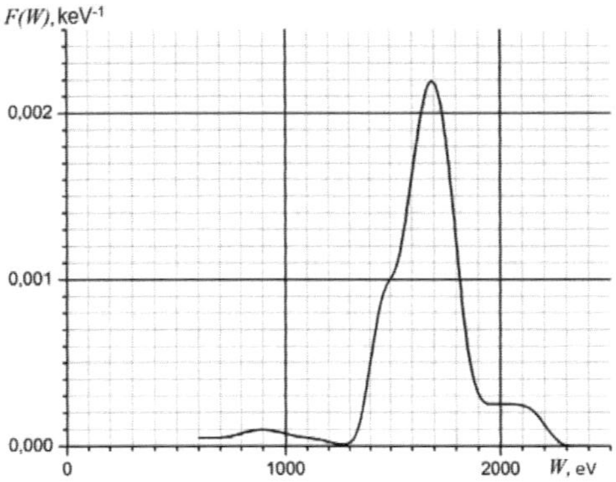

Figure 2. The kinetic energy distributions of ions. Discharge voltage is 2.4 kV at argon pressure of 10^{-3} Torr.

The algorithm of recovering the kinetic energy distributions of ions is given in [2]. Using the results of [2] for the three-electrode probe measurements we obtained the energy distributions of ions (figure 2).

The assembly of Faraday cups, which almost completely covers the cross-section of ion flow, was used to measure the integral current of ions emitted by the source and to study the radial distribution of current density (i.e. in the plane of cross-section perpendicular to the ion flow). Current density was determined by dividing the current magnitude by the area of the input aperture of the Faraday cup.

The experimental setup for measuring the emittance of corpuscular beam at the outlet of the ion source (the emittance meter) is shown in figure 3.

An analysis of the obtained emittance diagrams (figure 4) enables us to conclude that at low acceleration voltages the beam extracted from the ion source is not focused whereas the further increase of the extracting voltage makes the beam an axial one.

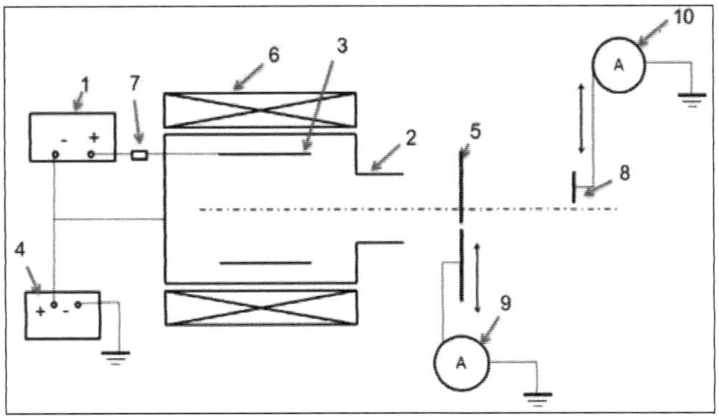

Figure 3. Scheme of experimental setup for emittance measurements: 1 – ion source's power supply; 2 – ion source's cathode; 3 – ion source's anode; 4 – power supply for positive biasing of the ion source with respect to the ground ; 5 – sliding plate with the round diaphragm; 6 – ion source's magnetic system; 7 – ballast resistance; 8 – sliding measuring probe; 9, 10 - ampermeters.

Journal of Physics: Conference Series **666** (2016) 012023 doi:10.1088/1742-6596/666/1/012023

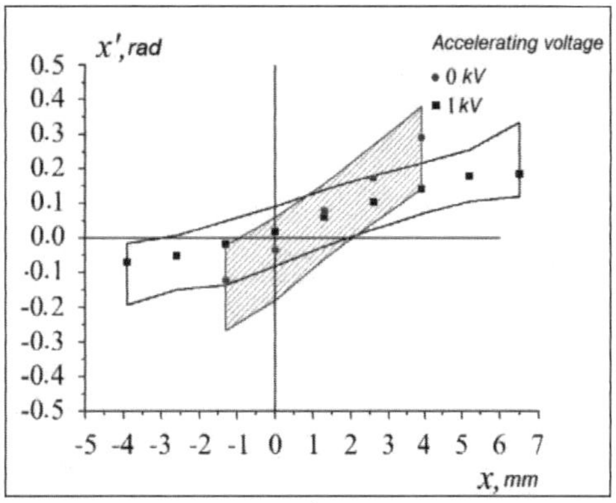

a)

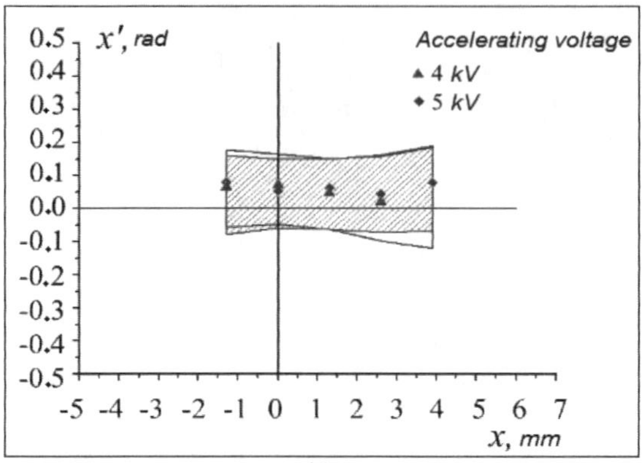

b)

Figure 4. Emittance diagrams of corpuscular beam extracted from ion source at different acceleration voltages: **a)** U_{acc} = 0 kV and 1 kV; **b)** U_{acc} = 4 kV and 5 kV.

The “glowing ion beams” method [3], based on the visualization of the ion beam’s trajectory due to glowing of residual gas in the visible light range of spectrum, was implemented. The trajectories were registered using a camera. The results of imaging (figure 5) enabled us to identify the improvement of the focusing properties of the accelerating gap with increasing acceleration voltage.

To simulate the motion of particle beam in the accelerating gap a 2.5D relativistic electromagnetic particle-in-cell (PIC) code [4] was used. Figure 6 shows the results of computer simulations of deuteron beam motion in the accelerating gap of the neutron tube. The results correlate with the experimental results shown in figure 5.

a)

b)

Figure 5. Photographic image of the ion beam in the accelerating gap at various acceleration voltages: **a)** U_{acc} = 0 kV; **б)** U_{acc} = -60 kV. 1 – low-voltage electrode of the accelerating gap from the side of the ion source (ground potential), 2 – high-voltage electrode of the accelerating.

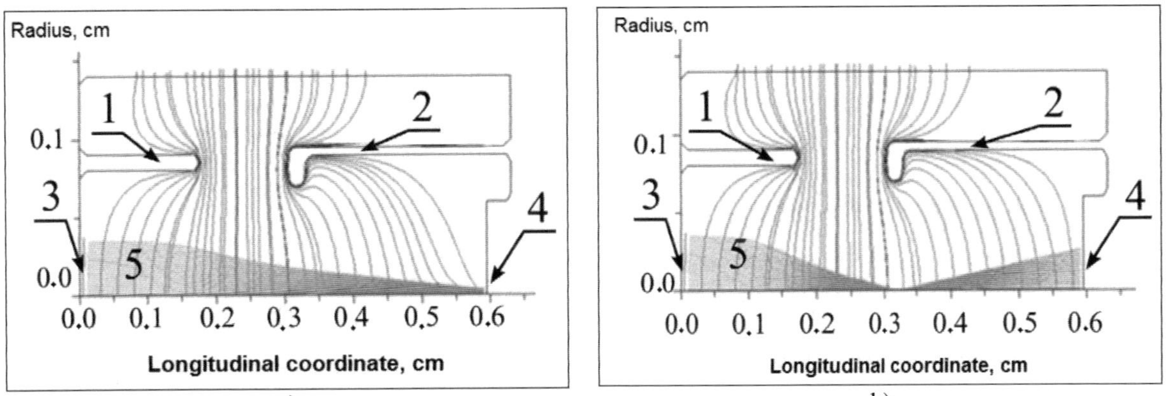

a) b)

Figure 6. Projections of equipotential surfaces of electric field on the imaging plane, and the shape of the transported ion beam in the accelerating gap, shown with grey background, at various acceleration voltages: **a)** U_{acc} = -10 kV; **b)** U_{acc} = -50 kV. 1 – low-voltage electrode from the side of the ion source, 2 – high-voltage electrode of the accelerating gap from the side of the target, 3 – outlet of the ion source, 4 – the plane of target's surface, 5 – the transported ion beam.

This work was supported by the grants of the Russian Foundation for Basic Research.

References

[1] Lochte-Holtgreven W 1968 *Plasma Diagnostics* ed North-Holland (Amsterdam: Interscience (Wiley) New York)

[2] Markov V G, Prohorovich D E, Sadilkin A G and Shhitov N N 2013 Determination of the corpuscular emission energy characteristics for the ion sources of gas-filled neutron tubes *Advances in Applied Physics* **1** 23–29 (in Russian)

[3] Belykh S F et al. 1992 *Tech. Phys.* **62** 179

[4] Rashhikov V I 1991 Space–Charge Field of the Beam in the Structure with Drift Tubes *Linear Accelerators* (Moscow: Energoatomizdat) 48–55 (in Russian)

Journal of Physics: Conference Series **666** (2016) 012024 doi:10.1088/1742-6596/666/1/012024

Formation of low-frequency periodic structures in a pulsed magnetron discharge

A V Kaziev[1], **G V Khodachenko**[2] **and M M Kharkov**
Plasma Physics Department, National Research Nuclear University MEPhI (Moscow Engineering Physics Institute), 31 Kashirskoe shosse, 115409 Moscow, Russia

E-mail: avkaziev@gmail.com

Abstract. Periodic plasma structures are observed in non-sputtering magnetron discharge (NSMD) that is the transient quasi-stationary low-voltage regime between the high-current magnetron discharge (HCIMD) and an arc. The fast camera imaging synchronized with the magnetic probe diagnostics reveals the correlation between the observed rotation of the plasma inhomogeneities and the magnetic field perturbation behaviour. The frequencies of the periodic processes fall into kHz-range. A simple analytical model of the ionization instability in crossed electric and magnetic fields is suggested for the low-pressure discharge case. Using the model, the possible ranges of wavelengths and frequencies for the plasma inhomogeneities are evaluated. The results show good agreement between the experimental data and theory.

1. Introduction

Recently, several research groups have reported observation and investigation of various wave-like phenomena in high-power impulse magnetron sputtering (HiPIMS) discharges (e.g. [1–3]). Considerable effort is being made to determine the mechanisms of drifting ionization zones appearing in such plasmas, and to gain control over their formation. Our present contribution deals with the investigations of similar effects taking place in quasi-stationary magnetron discharges: high-current impulse magnetron discharge (HCIMD), and low-pressure high-current impulse diffuse discharge (HCIDD), also known as non-sputtering magnetron discharge (NSMD) [4, 5]. HCIMD is a high-voltage (up to 1.5 kV), high-current (up to 20 A/cm^2), long (~ 1–40 ms) impulse sputtering tool [4]. The low-voltage (~ 80 V) NSMD dramatically inhibits the sputtering rate thus limiting the deposition process under certain conditions.

Here we consider the experimentally observed periodic structures in case of high-current impulse magnetron discharge on an aluminum target in Ar/CO_2 working gas mixture, and its transition into low-voltage non-sputtering magnetron discharge.

2. Experimental setup

The experiments were carried out in the conventional planar magnetron device as well as in the so-called apparatus with profiled electrodes. The scheme of the experimental setup and the diagnostic instruments is shown in figure 1.

---

[1] To whom any correspondence should be addressed.

[2] Deceased February 20, 2015.

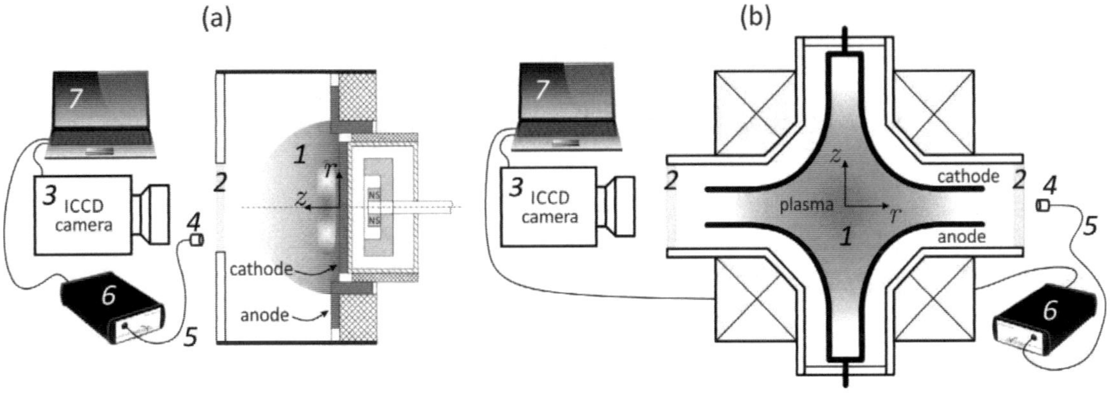

Figure 1. Experimental apparatus: (a) planar magnetron, (b) device with profiled electrodes: *1*—plasma, *2*—quartz window, *3*—fast camera, *4*—collimating lens, *5*—fiber optic, *6*—spectrometer, *7*—computer.

The planar magnetron consists of an 80-mm-diameter disk cathode, grounded anode and the magnetic system. The device with profiled electrodes consists of two hollow axisymmetric Al electrodes arranged between two electromagnetic coils. Current in the coils flows in the opposite directions so that the magnetic field lines form a cusp. The profile of each electrode replicates the magnetic field line curvature, ensuring $\mathbf{E}\cdot\mathbf{B} \approx 0$ along the most part of their surface. Further apparatus details could be found elsewhere [4, 5].

The discharge power supply system comprises two units: stationary power supply and pulsed power supply. High voltage pulses are generated by commutation of a transmission line using a high-power switch. The energy stored in the transmission line is determined by its charging voltage (e.g. 2 kV corresponds to 3.8 kJ energy). A stationary 0.1 A magnetron discharge provides low density initial plasma.

In the experiments discussed in present contribution the working gas was a mixture of Ar and CO_2 with partial pressures $p_{Ar} = p_{CO_2} = 0.5$ Pa .

A fast nine-frame gated camera BIFO K011 is used to perform high-speed imaging of the processes running in the discharge chamber. This device is capable of taking nine images with programmable frame times (100 ns–100 μs) and frame delays (100 ns–100 μs) after triggering. The spatial resolution of the matrix is 340×340 pixel for each frame. The spectral sensitivity range is 400–800 nm. Camera was equipped with Zenitar-M 50 mm lens. Simultaneously the Avantes AvaSpec-2048x14 spectrometer was used to record the optical emission spectrum of the plasma thus monitoring the non-sputtering regimes. The electrical parameters of the discharge (current I_d and voltage U_d) were recorded by the Tektronix TPS 2024 digital storage oscilloscope. The typical traces of transition from HCIMD to NSMD and then to an arc could be found elsewhere [5].

The fast camera imaging was synchronized with the magnetic probe diagnostics. Magnetic probe was used to measure internal magnetic field perturbations in pulsed plasma. It was calibrated using Helmholtz coils beforehand.

3. Experimental results

It was found that the NSMD plasma tends to produce clear optical emission inhomogeneities that rotate along the azimuthal direction. The typical series of fast images made in planar magnetron is presented in figure 2.

The NSMD plasma behaviour was found to be the same in both magnetron devices. The typical FFT spectrum of the magnetic probe signal is presented in figure 3 along with the fast images taken in the apparatus with profiled electrodes (side-on view).

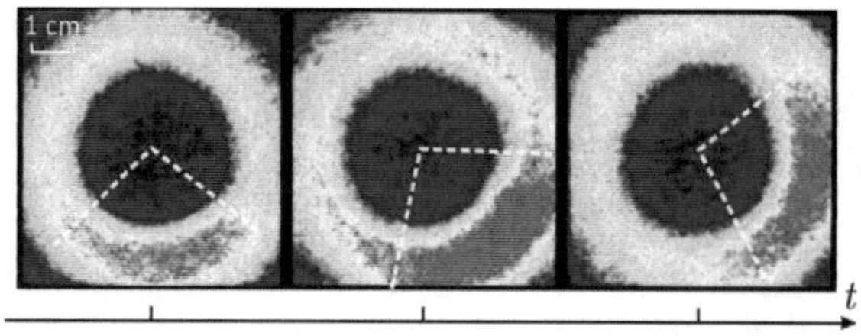

Figure 2. The fast images of NSMD plasma in planar magnetron (frame time 1 μs, frame delay 30 μs).

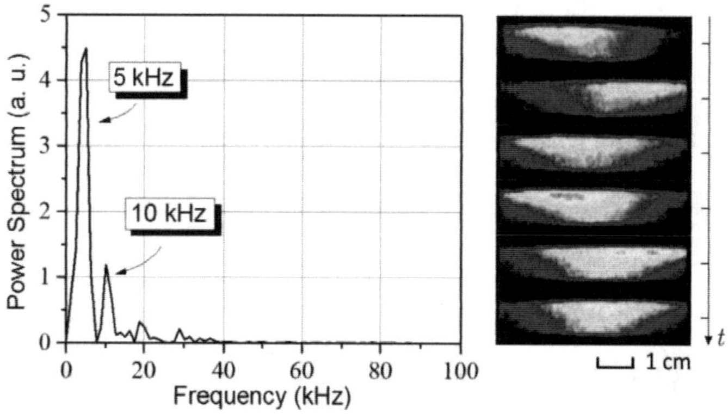

Figure 3. The low-frequency region of FFT spectrum of the magnetic probe signal (left) and the fast images of NSMD in apparatus with profiled electrodes (right) (frame time 1 μs, frame delay 100 μs).

It appears that the peaks in the FFT spectrum of magnetic probe signal correlate with the observed periodic processes in plasma. Namely, the rotation frequency of the bright regions was estimated as 2–10 kHz both from the fast camera imaging and magnetic probe measurements. Whenever the plasma perturbations were not observed, the low-frequency peaks vanished from the magnetic probe FFT spectra.

4. Model of the sheath processes

Dynamics of the comparable low-frequency processes occurring in Hall thrusters as well as in HiPIMS discharges has been well described in terms of predator-prey models [6, 7]. These models, however, do not consider the nature of the spatial and temporal scales of the azimuthal inhomogeneites. Hence, in order to get the satisfactory description of these phenomena in our experimental conditions we suggest a simple analytical model.

Because the observed periodic processes appear to feature the low frequencies ($f \sim$ 2–10 kHz), the proposed model is based on the classical considerations describing the formation of zones with enhanced ionization rate (striations) in the glow discharges [8].

We extend the existing model of the ionization instability [8] to the case of low-pressure plasma in crossed electric and magnetic fields. It is well known that the ionization instability may develop provided the ionization rate depends on the plasma density. Therefore this process is determined by the electron kinetics. Under the conditions typical for magnetron discharges, the cathode sheath is thin and non-collisional, while the inelastic processes that are essential for the discharge existence take place in **E**×**B** (azimuthal) direction. Thus in such plasmas the drift-diffusion approach is valid for the azimuthal direction only, and the particles' collisions don't play any role in the other directions. Let us replace a cylindrical system (seen in figure 1) with a Cartesian one and consider a near-cathode region where y-axis represents coordinate along the azimuthal direction in a magnetron device. Moreover, let the transverse magnetic field **B** be uniform so that electric and magnetic fields are orthogonal in the whole region considered. Note, that hence the following model is valid for the racetrack zone only. The model of the processes that take place in this simplified near-cathode region is shown in figure 4.

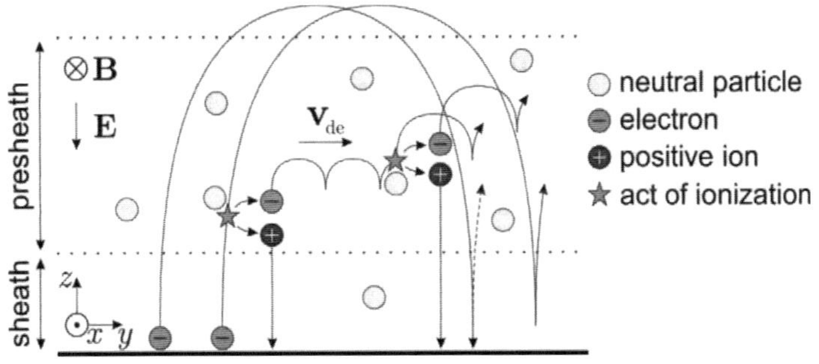

Figure 4. Main elementary processes in the near-cathode region of NSMD.

Here the electrons are emitted from the cathode surface under ion bombardment. Some of them cross the sheath and presheath regions without collisions and may return to the cathode. Those electrons that ionize atoms and molecules in the presheath generate slow secondary electrons that drift in crossed **E**×**B** field.

Like in a glow discharge case, here the electron drift current is considered as an external one with null divergence. The equation of charged particles balance in plasma appears to be:

$$\frac{\partial n}{\partial t} - D_{\mathrm{a}}^{(\mathrm{B})}\frac{\partial^2 n}{\partial y^2} = n\nu_{\mathrm{i}}, \tag{1}$$

where $D_{\mathrm{a}}^{(\mathrm{B})} = \left(b_{\mathrm{e}\perp}D_{\mathrm{i}\perp} + b_{\mathrm{i}\perp}D_{\mathrm{e}\perp}\right)/\left(b_{\mathrm{e}\perp} + b_{\mathrm{i}\perp}\right)$ is the ambipolar diffusion coefficient across the magnetic field lines [8] ($b_{\mathrm{e,\,i}}$ — electron and ion mobilities, $D_{\mathrm{e,\,i}}$ — their diffusion coefficients).

Let us suppose that there's a small density perturbation in plasma that is periodic along y axis. Assuming the simple harmonic character of this perturbation the plasma density becomes [8]:

$$n(x,y,z,t) = n^{(0)}(x,z) + n^{(1)}(x,y,z,t) = n^{(0)}(x,z) + \mathrm{Re}\left[n_k^{(1)}(x,z,t)\exp(-iky)\right], \tag{2}$$

where $n^{(0)}$ is the initial (unperturbed) plasma density, $n^{(1)}$ is the small perturbation, k determines the spatial scale or the wavelength of the perturbation: $\lambda_k = 2\pi/k$.

The instability develops if the growth rate $\gamma > 0$, i.e.

$$\gamma = n\frac{\partial \nu_{\mathrm{i}}}{\partial n} - k^2 D_{\mathrm{a}}^{(\mathrm{B})} > 0. \tag{3}$$

Taking into account that $\left(n/\nu_{\mathrm{i}}\right)\partial\nu_{\mathrm{i}}/\partial n \sim 1$, we find that if the frequency of ionization ν_{i} is larger than that associated to ambipolar diffusion ($k^2 D_{\mathrm{a}}^{(\mathrm{B})}$) the plasma density will grow. $D_{\mathrm{a}}^{(\mathrm{B})}$ and ν_{i} may be evaluated using the experimental data. Then the minimal spatial scale of the density perturbation in our model is limited by

$$\lambda_k^{\min} = 2\pi\sqrt{\frac{D_{\mathrm{a}}^{(\mathrm{B})}}{\nu_{\mathrm{i}}}}, \quad \lambda_k^{\min} = 0.5\text{–}5\ \mathrm{cm} \tag{4}$$

depending on the plasma parameters. The maximal λ_k value can be found after introducing the electron temperature perturbation. Considering the heating of electrons and their energy balance the relation between the electric field and the electron temperature T_{e} can be found [8]. Then taking into account the simple harmonic electron temperature perturbation and its relation to the density fluctuation, the balance of charged particles (1) turns into [8]:

$$\frac{\partial n_k^{(1)}}{\partial t} = \mathrm{i}\frac{\sqrt{6}}{5}\frac{T_e}{k\lambda_{T_e}}\frac{\partial \nu_i}{\partial T_e} n_k^{(1)} + \left(n\frac{\partial \nu_i}{\partial n} - D_a^{(B)} k^2 - \frac{6}{5}\frac{T_e}{k^2\lambda_{T_e}^2}\frac{\partial \nu_i}{\partial T_e} \right) n_k^{(1)}, \tag{5}$$

where λ_{T_e} is the length of electron temperature relaxation, i.e. the path the electron travels until it loses kinetic energy equal to T_e due to collisions with other particles (we express T_e in the energy units):

$$\lambda_{T_e} = \sqrt{\frac{b_e T_e}{e}\frac{1}{\kappa_{eg}\nu_{eg}}}, \tag{6}$$

where κ_{eg} is the mean fraction of electron energy lost in collisions with neutral gas particles, ν_{eg} — frequency of such collisions [8].

In our case the electrons produce a current along y axis with current density:

$$j = env_d = en\frac{E}{B}. \tag{7}$$

So for the magnetized electrons the modified mobility becomes $b_e = 1/B$ and hence the temperature relaxation happens on a scale of

$$\lambda_{T_e} = \sqrt{\frac{T_e}{eB}\frac{1}{\kappa_{eg}\nu_{eg}}}, \tag{8}$$

Solving (6) we derive $n_k^{(1)}$ and $n^{(1)}$:

$$n^{(1)} = \mathrm{Re}\left[n_k^{(1)} \exp(-iky) \right] = n\exp(\gamma t)\cos(\omega t - ky),$$
$$\omega = \frac{\sqrt{6}}{5}\frac{T_e}{k\lambda_{T_e}}\frac{\partial \nu_i}{\partial T_e}, \tag{9}$$
$$\gamma = n\frac{\partial \nu_i}{\partial n} - D_a^{(B)}k^2 - \frac{6}{5}\frac{T_e}{k^2\lambda_{T_e}^2}\frac{\partial \nu_i}{\partial T_e}.$$

The upper limit of the spatial scale of perturbation can be found from

$$\gamma > 0 \quad \Leftrightarrow \quad k^2\lambda_{T_e}^2 > \left(\frac{T_e}{\nu_i}\frac{\partial \nu_i}{\partial T_e} \right)\left(\frac{n}{\nu_i}\frac{\partial \nu_i}{\partial n} \right)^{-1}. \tag{10}$$

Eventually the maximal wavelength is

$$\lambda_k^{\max} = 2\pi\lambda_{T_e}, \quad \lambda_k^{\max} \sim 100\ \text{cm} \tag{11}$$

Thus the maximal perturbation wavelength is limited by the length of electron temperature relaxation. The latter strongly depends on the prevailing electron energy loss mechanism. Generally speaking, $\kappa_{eg}\nu_{eg} = \kappa_{eg}^{(el)}\nu_{eg}^{(el)} + \kappa_{eg}^{(inel)}\nu_{eg}^{(inel)}$, where superscripts (el) and (inel) correspond to elastic and inelastic processes. So the electrons with energies lower than the inelastic processes threshold mostly lose their energy in elastic collisions with atoms and molecules. On the contrary, electrons with rather high energies sufficient for excitation and/or ionization processes predominantly lose their energy in inelastic collisions. Nevertheless their frequency is lower, the fraction of energy lost is much higher for inelastic processes.

Then the situation may be clarified if we evaluate the energy electrons gain on their cycloidal paths: $\varepsilon_{ec} = eE_\perp h = 2m_e E_\perp^2 / B^2$. The internal electric field in the presheath is experimentally measured: $E_\perp \sim 10^4\ \text{V/m}$. The magnetic field decreases from 0.03 T to 0.01 T when moving away from the cathode surface across the sheath and presheath.

Then the electrons drifting close to the cathode surface (in the lower region in figure 2) gain $\varepsilon_{ec} \sim 1\,\text{eV}$. In this case the elastic losses dominate, hence $\kappa_{eg}\nu_{eg} \approx \kappa_{eg}^{(el)}\nu_{eg}^{(el)}$ and $\lambda_k^{max} = 2\pi\lambda_{T_e} \sim 100\,\text{cm}$.

On the contrary, electrons drifting in the presheath (in the upper region in figure 2) gain $\varepsilon_{ec} \sim 10\,\text{eV}$. Here $\kappa_{eg}\nu_{eg} \approx \kappa_{eg}^{(inel)}\nu_{eg}^{(inel)}$ and the maximal wavelength decreases: $\lambda_k^{max} = 2\pi\lambda_{T_e} \sim 10\,\text{cm}$.

Our evaluations of the energies are valid for the average values only. Obviously, both the elastic and inelastic processes take place in the whole presheath while the accurate $\kappa_{eg}\nu_{eg}$ value must be calculated using the electron energy distribution function. Nevertheless, our approach allows determining upper and lower limits for the ionization instability wavelength.

The spatial parameters of the experimentally observed plasma inhomogeneities match those evaluated using the simple model, i.e. 1 cm < λ << 100 cm. The frequency of the perturbation wave according to (9) gives $f = \omega/(2\pi) = 10^2$–10^4 Hz. Thus both the spatial and time scales of the observed wave phenomena in NSMD are in good agreement with the theoretical results. The proposed analytical model may be also adopted for describing the plasma dynamics in high-power sputtering magnetron modes (HCIMD and HiPIMS).

5. Conclusion

The low-frequency (2–10 kHz) periodic structures are observed and investigated in non-sputtering magnetron discharge operated on Al target in oxygen-containing working gas mixture. A simple analytical model of the ionization instability in crossed electric and magnetic fields is proposed. Experimental results are in good agreement with theory.

Acknowledgment

We thank Prof. Yury A. Lebedev (Topchiev Institute of Petrochemical Synthesis, RAS) for providing the fast camera for our experiments.

The work was produced during the joint project between USATU (Ufa State Aviation Technical University) and UMPO (Ufa Engine Industrial Association) with title “Elaboration and industrial development of high-precision shaping coordinated technologies and superficial hardening of responsible details from Al-alloys with heightened constructional energy efficiency”, implemented under the contract №40/10-30976/НЧ-НЧ-01-13-ХГ sponsored by the Ministry of Education and Science of the Russian Federation (contract №02.G25.31.0010 between UMPO and the Ministry of Education and Science of the Russian Federation) through the Resolution of the Russian Federation Government № 218 from April 9, 2010.

References

[1] Kozyrev A V, Sochugov N S, Oskomov K V, Zakharov A N and Odivanova A N 2011 *Plasma Phys. Rep.* **37** 621

[2] Winter J, Hecimovic A, de los Arcos T, Böke M and Schulz-von der Gathen V 2013 *J. Phys.* D: *Appl. Phys.* **46** 084007

[3] Anders A, Ni P and Rauch A 2012 *J. Appl. Phys.* **111** 053304

[4] Khodachenko G V, Mozgrin D V, Fetisov I K and Stepanova T V 2012 *Plasma Phys. Rep.* **38** 78

[5] Kaziev A V and Khodachenko G V 2014 *Phys. Scr.* **T161** 014053

[6] Gallian S, Hitchon W N G, Eremin D, Mussenbrock T and Brinkmann R P 2013 *Plasma Sources Sci. Technol.* **22** 055012

[7] Barral S and Ahedo E 2008 *AIP CP* **993** “Plasma 2007” 439

[8] Golant V E, Zhilinsky A P and Sakharov I E 1980 *Fundamentals of Plasma Physics* (New York: John Wiley & Sons)

Simulation and experimental investigation of hydrogen target plasma parameters for experiments on heavy ion beams deceleration

A P Kuznetsov[1,2,4], K L Gubskiy[2], O A Byalkovskiy[3], R O Gavrilin[1], A A Golubev[1,2], I V Rudskoy[1], S M Savin[1], D S Koshkin[2], A M Korolev[2], A.V. Mikhaylyuk[2] and A O Khurchiev[2]
[1] Institute of Theoretical and Experimental Physics SSC RF, Bol'shaya Cheremushkinskaya ul. 25, 117218 Russia
[2] National Research Nuclear University MEPhI, Kashirskoe sh. 31, Moscow, 115409 Russian Federation
[3] IRE–Polus Co., Fryazino, Moscow oblast, Russia

E-mail: apkuznetsov@mephi.ru

Abstract. For the purposes of the experiments on heavy ions deceleration in ionized matter a plasma target based on a linear electric discharge in hydrogen has been designed. The two-dimensional hydrodynamic code was used for numeric simulation of the main stage of discharge. Estimated distribution of electron density and the degree of plasma ionization along the discharge tube axis of the target at the time of maximum discharge current corresponds to the experimental measurements of time dynamic of linear electron density and degree of plasma ionization in the target.

1. Introduction
Issues of charged particle beams energy losses in a cold matter and plasmas have a variety of practical applications: in the design of inertial confinement fusion targets, in material science, medicine, acceleration physics. Generally, energy losses of a charged particle in matter are caused not only by elastic and inelastic nuclear collisions, but also by the Coulomb interaction of a particle with target atoms [1].

In order to study the processes that take place with ions deceleration in an ionized matter, plasma targets, created by external energy sources such as a discharge in gas [2,3], capillary discharge [4], laser plasma [5,6], explosive generator plasma [7], and inserted into ion beam transportation line, are widely used.

The ion energy losses are difficult to estimate for a number of reasons. Calculation of ion charge variation dynamics requires the data for ionization and recombination cross sections. In the case of deceleration of ions with low energies in plasma, the required cross sections are practically unknown, and the results of estimation using different theoretical models may vary by an order of magnitude. Therefore it is relevant to perform direct measurements of the ion energy losses both in a "cold" and an ionized matter. One of the key aspects of these experiments is revealing the correlation between the ion energy losses and the ionized matter parameters: density, temperature and the degree of ionization.

The results of numerical simulation and experimental measurements of hydrogen target plasma electron density and the degree of ionization are presented in this article.

2. Numeric simulation of hydrogen target plasma parameters

At the Institute of Theoretical and Experimental Physics, a plasma target based on a linear electric discharge in hydrogen was designed for experiments on the charged particle beams deceleration in an ionized matter. The target plasma is obtained by an electric discharge in two 78-mm-long collinear quartz tubes of an inner diameter of 6 mm (figure 1). A 3-μF capacitor battery charged to 2–5 kV initiates a current of up to 400 A in each discharge channel. The currents in the discharge channels have opposite directions, due to which the effect of ion beam focusing, caused by the magnetic field of the discharge current decreases substantially. The vacuum system of the ion transportation line is protected from the gas load of the target by means of differential evacuation with the use of a turbo-molecular pump with an evacuation rate of 500 L/s. In this case, the length of the gas column in the target (l = 220 mm) is limited by two diaphragms with 1-mm-diameter apertures. The oscillograms of the current (1) and discharge luminosity (2) in the spectral range of 400 – 1100 nm, obtained at the charge voltage U=5 kV and hydrogen pressure p=1 mbar, are shown in figure 1b.

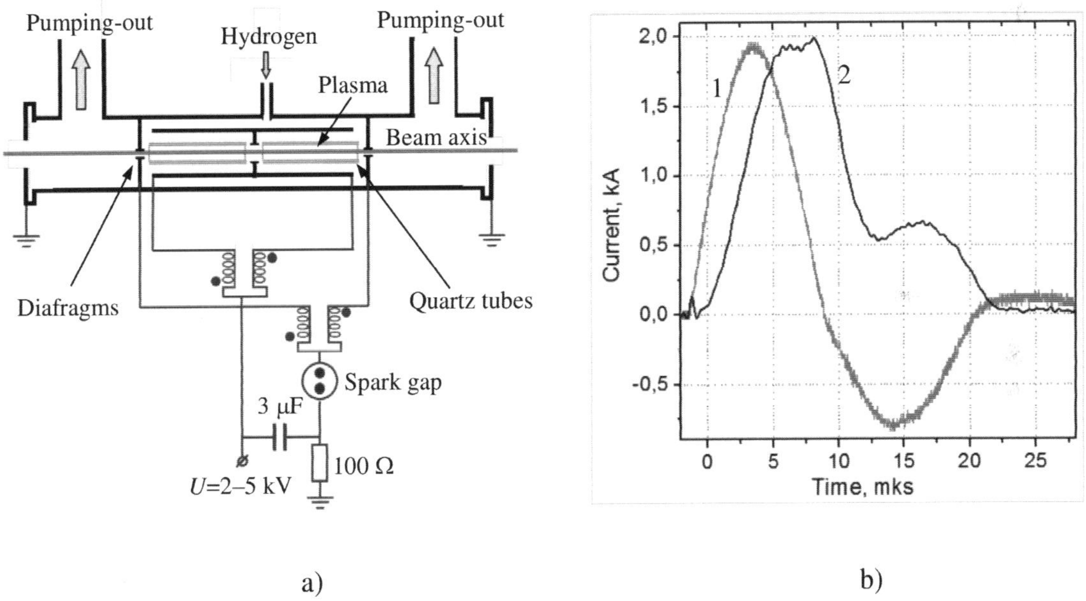

Figure 1. Schematic of a hydrogen plasma target (a) and the current (1) and discharge luminosity (2) oscillograms (b).

To perform numeric simulation of the discharge stage, where the electric fields in the tube are already formed (~0.5 μs after the discharge initiation and later), the two-dimensional hydrodynamic code was used [9]. The ohmic gas heating, thermal conductivity and radiation energy losses were taken into account in the calculations. The ionization and recombination processes were considered using the average ion charge approximation. The discharge current was derived from the external electrical circuit differential equation. It was shown that radiation losses caused by bound-bound transitions appear to be the main channel of plasma energy losses. The resonant radiation is substantially absorbed inside the target while the plasma remains transparent for the other radiation transitions. The radiation losses could not be derived using this approach, therefore the model of the black body with the grayness factor ε as a parameter was applied. Estimated discharge current versus time at different ε are shown in figure 2. As can be seen, theoretical estimation best fits the experimental results when ε = 0.01. The results of temperature distribution, electron density and plasma ionization degree along the target discharge tube axis at the time instant t = 3.5 μs, which matches the discharge current maximum, are presented in figures 3a-c.

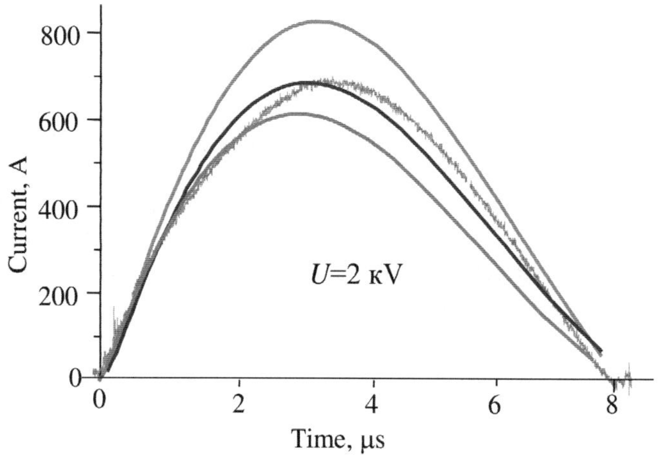

Figure 2. The results of the hydrogen target plasma parameters simulation: time dependence of the discharge parameters, experimental (red) and estimated for ε = 0.0001 (green), ε = 0.01 (blue), ε = 0.1 (orange), is presented.

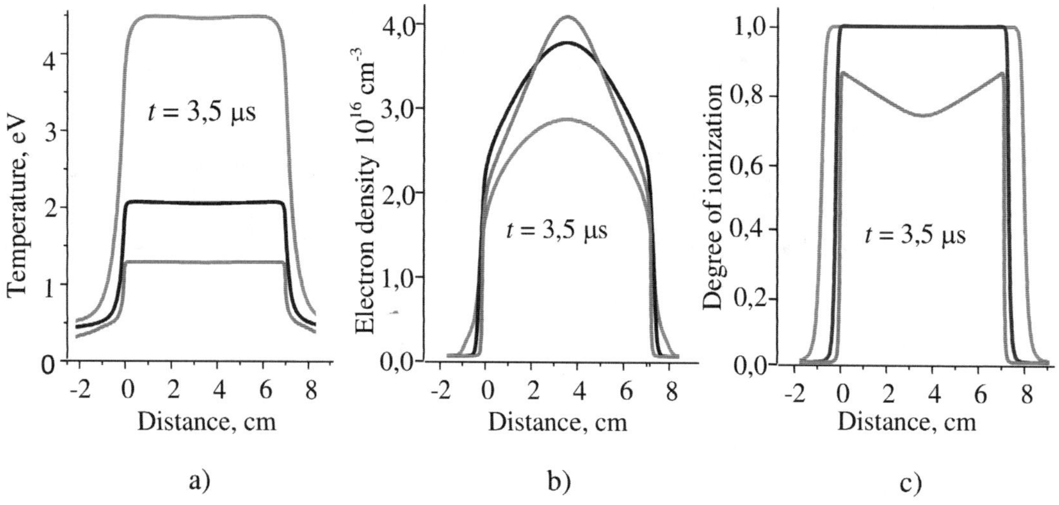

Figure 3. The results of simulating the hydrogen target plasma parameters: distributions of temperature (a), electron density (b) and the degree of plasma ionization (c) along the discharge tube axis of the target. Estimations for ε = 0,0001 (green), ε = 0,01 (blue), ε = 0,1 (orange) are presented.

3. Measurement of electron density and degree of plasma ionization

Among all the nonintrusive methods of plasma diagnostic the most accurate information about electron density *n* can be derived by means of interferometric techniques. In a multicomponent plasma with a low degree of ionization, the contributions from different kinds of particles (electrons, ions in different charge states, and neutral atoms) to the refractive index can be considered additive.

In gases at temperatures below 10 eV, the maximum contribution to n is made by neutral atoms in the ground state, and the refraction in the visible light spectral region is practically independent of the wavelength. In this case, the value of the phase shift δ of the probing electromagnetic wave is determined by the additive contribution of free electrons and variations in the neutral particle density caused by hydrodynamic processes in the heated gas. Due to strong plasma dispersion, it is possible to separate the contribution made by electrons to the phase shift of the probing electromagnetic wave from that made by heavy particles. To this end, it is necessary to perform the interferometric measurements simultaneously at two wavelengths.

Since the atom of hydrogen can exist in only two charge states – either neutral or fully ionized, it is possible to find the linear densities of electrons and hydrogen atoms N_H in hydrogen plasma from the phase shifts of the probe electromagnetic wave, measured simultaneously at two wavelengths [10].

To investigate the temporal dynamic of electron density and the degree of plasma ionization a two-wavelength interferometer with the quadrature signal registration was designed previously [11]. The 25 mW He-Ne laser at $\lambda = 632.8$ nm and the 50 mW single-frequency solid-state $YVO_4:Nd^{3+}$ laser with intracavity second harmonic generation ($\lambda = 532$ nm) were used as the sources.

The linear density of free electrons and the degree of plasma ionization in a hydrogen target were measured for initial hydrogen pressures of 1–8 mbar and capacitor battery voltages of 2–5 kV. The linear electron density [cm^{-2}] and the degree of ionization dependence on the initial hydrogen pressure at the maximum discharge current time are presented in figure 4.

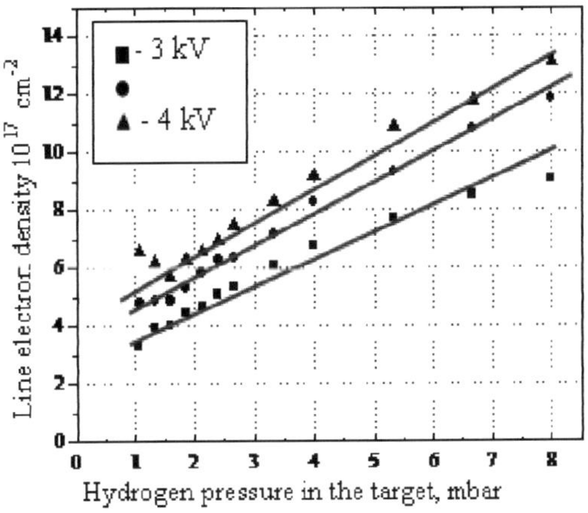

a)

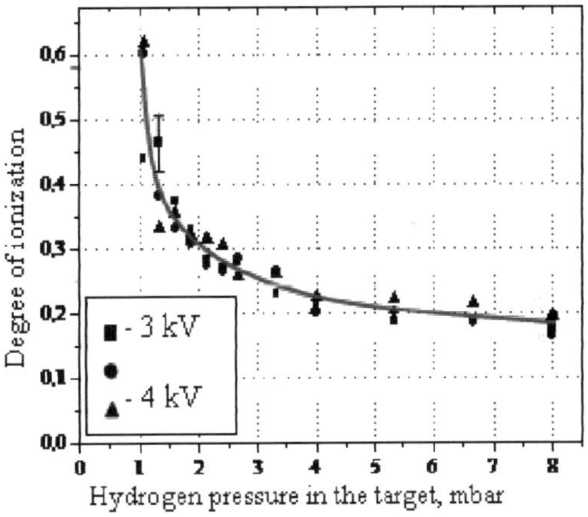

b)

Figure 4. Linear electron plasma density (a) and the degree of plasma ionization (b) at the maximum of the discharge current as functions of the initial hydrogen pressure for different values of the capacitor battery voltage.

As shown in figure 4a, with the voltage U and the initial gas pressure increase, the linear electron density increases almost linearly and reaches its maximum of $(1.31 \pm 0.01)\cdot 10^{18}$ cm^{-2} at the voltage of 5 kV and pressure of 8 mbar. As was expected, the maximum degree of hydrogen ionization is attained at the minimal amount of the plasma-forming gas (figure 4b). Within the experimental error, the degree of ionization at initial hydrogen pressures of 1–8 mbar is almost independent of the capacitor battery voltage U. The maximum degree of ionization 0.62 ± 0.05 was achieved at an initial pressure of 1 mbar.

4. Conclusions

Estimated distribution of the electron density and the degree of plasma ionization along the discharge tube axis of the target at the time of maximum discharge current corresponds to experimental measurements of time dynamics of the linear electron density and the degree of plasma ionization in the target. The measurements proved that the linear electron density can be varied in the range from $3.3 \cdot 10^{17}$ to $1.3 \cdot 10^{18}$ cm^{-2} by varying the discharge current and the initial hydrogen pressure. The maximum degree of plasma ionization 0.62 ± 0.05 was achieved at the initial pressure of 1 mbar.

References

[1] Deutsch C et al. 2010 *The Open Plasma Physics Journal* **3** 88-115
[2] Hofmann D H H et al. 1990 *Phys. Rev. A.* **42** 2313
[3] Koshkarev D G 2002 *Laser and particle beams* **20** 595
[4] Golubev A et al. 2001 *Nucl. Instrum. Meth. Phys. Res. A.* **464** 247
[5] Iwase O et al. 2000 *Europhys. Lett.* **50** 28
[6] Faik S, Tauschwitza A, Basko M M, Maruhna J A, Rosmej O, Rieneckera T, Novikov V G and Grushin A S 2014 *High Energy Density Physics* **10** 47
[7] Mintsev V et al. 1999 *Contributions to Plasma Physics* **39** 45
[8] Belyaev G et al 1996 *Phys. Rev. E.* **53** 2701
[9] Roudskoy I V 1996 *Laser and Particle Beams* **14** 369
[10] Kuznetsov A P, Golubev A A, Kozin G I, Mutin T Yu, Savelov A S and Fertman A D 2006 *Instruments and Experimental Techniques* **49** 247
[11] Kuznetsov A P, Byalkovskii O A, Gavrilin R O, Golubev A A, Gubskii K L, Rudskoi I V, Savin S M, Turtikov V I and Khudomyasov A V 2013 *Plasma Physics Reports* **39** 248

Journal of Physics: Conference Series **666** (2016) 012026 doi:10.1088/1742-6596/666/1/012026

Micro-particle charge determination using a linear Paul trap with the end electrode

D S Lapitsky[1]
Joint Institute for High Temperatures of the Russian Academy of Sciences (JIHT RAS), 125412, Izhorskaya st. 13 Bd.2, Moscow, Russia

E-mail: dmitrucho@yandex.ru

Abstract. A method for measuring the charge-to-mass ratio of micron-sized particles, using an electrodynamic linear Paul trap supplemented with the end electrode, is presented. Measurements were performed for the particle charge acquired by passing through a corona discharge.

1. Introduction

Measurement of the electrical charge of a single dust particle is important for technologies of dust removal from the air or from the industrial gas emissions, for ore enrichment using electric field, for studies of lightning and dust plasmas and for other applications.

Diverse methods of micro-particles charge-to-mass ratio measurement, i.e. the ratio of the total charge of particles to their total mass, are currently available [1]. The electrical charge on the particles can be measured by high-resistance electrometers, by the induction method or by analyzing the particle motion in an electric field. The induction method employs measuring the induction current occurring in a ring electrode when the charged particles pass through it [2].

The simplest method is determination of the charge by deviation of the particle track in an electric field. Originally this method was used in experiments of Millikan [3]. It was developed further by Laby and Hopper in Ref. [4] where they studied the motion of a charged oil droplet in a static electric field of a flat capacitor. The advantage of this method is the possibility of determining the particle size by its inhibition in the gaseous medium. This method is applicable for a constant, static electric field and for the variable field.

In a rarified medium, measurement of the ratio of charge to mass with alternating electric fields is used for the selective capture of ions and molecules. Approach based on trapping the charged particles in alternating electric fields in a gaseous medium at atmospheric pressure is suggested in [5], where particles are trapped by an alternating electric field formed by the Paul ion trap [6]. Due to the dissipation of the particle energy in a gaseous medium, the charged particle confinement range is wider than in the case of the rarified media [7, 8]. By studying the response of the particles to changes of the electric field parameters inside the trap (AC vs. DC electric fields and the frequency of the alternating electric field), it is possible to determine the ratio of the charge to the mass of the particles in this type of trap.

However, the use of the Paul ion trap for determination of the charge-to-mass ratio of the particles brings forward a number of challenges, such as those associated with long-term study of the response

[1] To whom any correspondence should be addressed.

of the amplitude of the particle oscillations to a change of the amplitude of the variable voltage or of the constant electrostatic potential [5].

In this paper a new method of measuring the charge-to-mass ratio of particles, using an electrodynamic linear Paul trap [6] supplemented with the end electrode, is proposed. It is shown that the charge-to-mass ratio of the particles pre-charged in a corona discharge can reach values of $6.8\cdot10^{-6}$ C/kg.

2. Method of particle charge measurement

The schematics of the experiment is shown in Figure. 1.

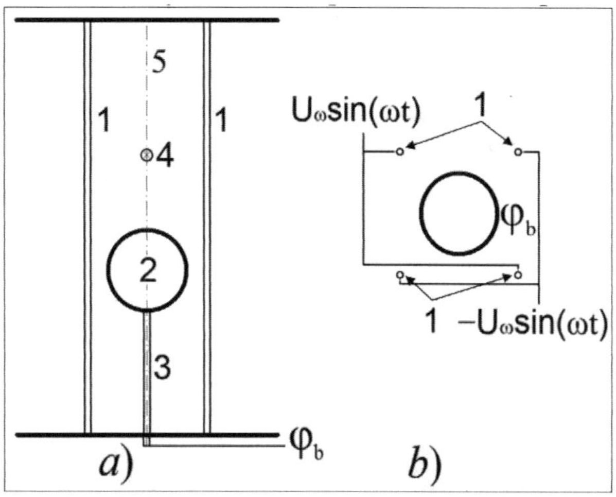

Figure1. The scheme of the particle charge-to-mass ratio measurement, a) – the side view, b) – the top view. 1 – electrodes of the Paul linear trap, 2 – ball-shaped electrode, 3 – thin rod, 4 – confined charged particle, 5 – central axis of the trap.

The AC voltage with amplitude U_ω is applied to the linear Paul trap electrodes (1). A charged particle in such an alternating electric field is confined to the area of the smallest amplitude of the electric field changes during the alternations. Inside the trap, this area corresponds to the central axis of the trap in the middle among the electrodes, and the particles will attract to this axis [5]. Contained in the horizontal plane, the charged particle (4) will move down along the axis of the trap under the force of gravity. To compensate the weight of the particles, the ball-shaped electrode (2) is installed on the central axis of the trap near the center of the trap on a thin rod (3). Particle repulsive potential U_b is applied to the ball-shaped electrode. Ball shape of the electrode is selected to reduce the non-uniformity of the electrostatic field that could be significant if using only the rod without the ball.

Knowing the distance between the ball-shaped electrode and the particle, one can estimate the particle charge-to-mass ratio from the balance between the gravity acting on the particle $F_g = m_p g$ and the Coulomb repulsion force $F_e = q_p E$, where m_p is the particle mass, g the gravity acceleration, q_p the particle charge and E the electric field.

3. Realisation of the method

Trap electrodes (1) (figure 1) were made of metal rods with a diameter of 4 mm. An alternating voltages $U_\omega\sin(\omega t)$ and $-U_\omega\sin(\omega t)$ were applied to the electrodes, U_ω = 840V, $\omega = 2\pi f$, f = 50Hz. On the central axis of the trap, at the lower end, a metal ball-shaped electrode of the radius R_b = 4 mm was installed on an 18 mm long base rod and the DC voltage U_b = 250, 500, 750 or 1000 V was applied to the ball-shaped electrode.

The particles were illuminated by a spread of the laser light and recorded by a high-speed HiSpec Fastec Imaging video camera with a maximum resolution of 1280x1024 pixels.

In the experiment, the dielectric powder particles of aluminium oxide Al_2O_3 with the average radius r_p = 17 microns, density ρ_p = 3990 mg/cm^3 and the dielectric constant ε = 10 were used.

Pre-charged in a corona discharge with the field strength of E = 7.7 kV/cm, the particles were injected into the trap, where they were captured by the trap field in the horizontal plane and by the interaction with the field of the ball-shaped electrode in the vertical direction. Figure 2 shows the ensemble of the suspended particles above the ball electrode in a form of a chain. By changing the U_b potential, the particles position above the electrode was varied.

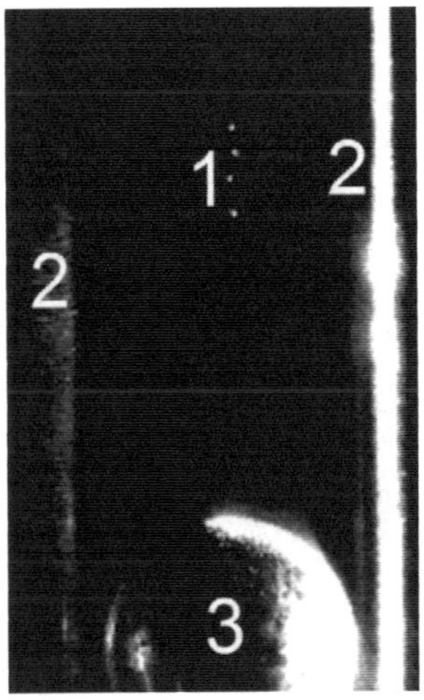

Figure 2. Particles confined inside the linear trap with the ball-shaped electrode. 1 – the charged particles above the ball-shaped electrode, 2 – the linear trap electrodes, 3 – the end ball-shaped electrode. The voltage U_b = 250 V forms the electric field similar to that of the point charge in the middle of the ball $Q_b = 1.14\cdot10^{-10}$ C.

Trap electrodes perturb the electric field of the ball-shaped electrode. The calculation of the effect of the trap electrodes on the vertical distribution of the ball-shaped electrode field along the axis of the trap was performed using the Comsol multiphysics software in the stationary case for the time instant $\omega t = \pi/2$. The electric potential distribution is shown in figure 3. This potential with an increase of the distance r from the ball decreases faster than $U(r) = q_b{\cdot}k/r$, where $k = 8.98\cdot10^9$ m/s, and can be approximated by a function $U(r) = q_b{\cdot}\, k/r^{2.34}$ (figure. 3).

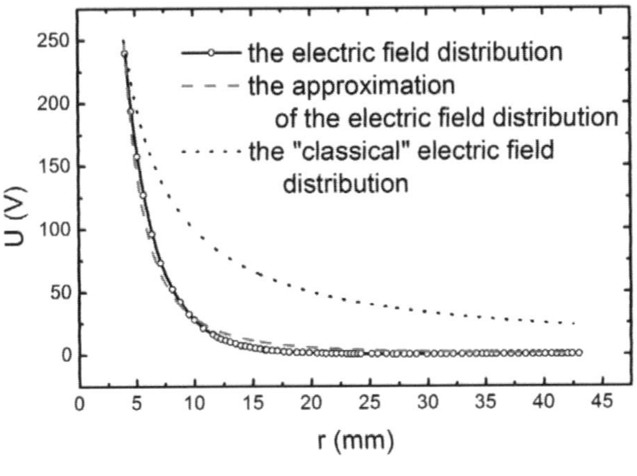

Figure 3. Comparison of the distribution of the electric field of the ball-shaped electrode with the trap electrode and without it. The end electrode potential U_b = 250 V.

Since in the confined structure several particles were captured and arranged in a vertical chain, the charge of the suspended particles was calculated for the top and bottom particles in the chain with the influence of the remaining particles treated under the assumption that they all have the same charge q_p. For each value of U_b (250, 500, 750, 1000 V), the distances between the particles and the distances between them and the end electrode (as in figure 2 at U_b = 250) were determined.

Then, the particle charge was calculated from the following formula:

$$m_p \vec{g} = q_p \overrightarrow{E(r_{p-b})} + \sum_i \vec{F}^{\,i}_{\text{int}} \quad (1)$$

where F^{i}_{int} is the Coulomb interaction force between the studied particle p and particle i, $r_{\text{p-b}}$ the distance from the particle p to the center of the ball-shaped electrode and $E(r_{\text{p-b}})$ the field strength at the distance $r_{\text{p-b}}$ ($E(r)=U(r)/r$). Considering the effect from the trap electrodes on the end electrode field distribution, the equation (1) can be re-written as:

$$m_p \vec{g} = k \frac{Q_b q_p}{\vec{r}_{p-b}^{\,3.34}} + \sum_i k \frac{q_p^2}{\vec{r}_{p-i}^{\,3.34}} \quad (2)$$

Here Q_{p} is the ball-shaped electrode charge corresponding to the potential applied to the electrode ($Q_p = R_b \cdot \varphi_b / k$) and $r_{\text{p-i}}$ the distance between the particle p and particle i.

The average charge-to-mass ratio was found to be $q_{\text{p}}/m_{\text{p}} = 6.8 \cdot 10^{-6}$ C/kg ($q_{\text{p}} = 3 \cdot 10^4$ e) with the mean-square deviation $\sigma = 1.34 \cdot 10^{-6}$ C/kg.

4. Conclusion

A new method is proposed for measuring the charge-to-mass ratio of the particles by their confinement in the Paul linear trap with the end electrode added and measuring the particle position over the end electrode.

Specific charge of the particles passing through the corona discharge field with the field strength $E = 7.7$ kV/cm was found to be $q_{\text{p}}/m_{\text{p}} = 6.8 \cdot 10^{-6}$ C/kg with mean-square deviation $\sigma = 1.34 \cdot 10^{-6}$ C/kg.

5. Acknowledgements

This work was conducted with the financial support of Russian Science Foundation via grant number 14-50-00124.

References

[1] Vereschagin I P, Levitov V I, Mirzabekian G Z and Pashin M M 1974 *Fundamentals of the electrogasdynamics in disperse systems* (Moscow: Energia) 160
[2] Gunn R 1947 *Phys. Rev.* PR3 **71**
[3] Millikan R A 1910 A new modification of the cloud method of determining the elementary electrical charge *Phil. Mag.* **19** 209
[4] Laby T H and Hopper V D 1939 *Nature* **143** 157
[5] Winter H, Ortjohann HW 1991 *Am. J. Phys.* **59** 807
[6] Paul. V. 1990 *UFN* **160** 112
[7] Lapitskiy D S, Filinov V S, Deputatova L V, Vasilyak L M, Vladimirov V I and Pecherkin V Ya 2015 *High Temperature* **53** 1
[8] Vasilyak L M, Vladimirov V I, Deputatova L V, Lapitsky D S, Molotkov V I, Pecherkin V Ya, Filinov V S and Fortov V E 2013 *New. J. Phys.* **15** 043047

Journal of Physics: Conference Series **666** (2016) 012027 doi:10.1088/1742-6596/666/1/012027

Surface diagnostics by ion scattering spectroscopy in gaseous environment

N V Mamedov[1,2,3], **V A Kurnaev**[2], **D N Sinelnikov**[2] **and D V Kolodko**[2]
[1] Federal State Unitary Enterprise «All-Russia Research Institute of Automatics», Sushchevskaya street 22, Moscow, 127055, Russian Federation
[2] National Research Nuclear University MEPhI, Kashirskoe shosse 31, Moscow, 115409, Russian Federation

E-Mail: m_nikitos@mail.ru

Abstract. Nowadays, it is important to perform *in-situ* analysis of composition and thickness of ultra-thin (~5-50 Å) surface layers in the course of the surface exposure to plasma or its components. For this aim, a new experimental facility based on the MEPhI Mass Monochromator is being developed, where low and medium energy ion spectroscopy of samples is used just after or during plasma exposure/ion treatment. A differentially pumped energy analyzer is used for recoil ion spectroscopy under grazing incidence conditions in the automated ion mass monochromator. A built-in Penning plasma source is used for plasma/ion treatment of samples. In this paper, the influence of the working gas (during the plasma source operation) on the energy spectra of reflected and recoil ions is studied. It is shown that the peak shape of the energy spectrum of scattered H^+ ions increases during gas injection, and, at the same time, the signal intensity of the ionized recoils from the sample decreases. Nevertheless, analysis of the surface composition and thickness of the outer layer could be done at pressures ranging up to 10^{-4} Torr.

1. Introduction

At present, understanding the dynamics of deposition and sputtering of the materials of the plasma-facing components (PFCs) in fusion devices is essential. In order to improve the plasma parameters, material migration and re/codeposition, which take place under plasma irradiation, are used in addition to boronization and litization of the PFCs.

Well-known linear plasma simulators (e.g. PISCES, NAGDIS, Pilot-PSI) are used to investigate the plasma-surface interaction. However, the plasma-exposed surfaces are typically analyzed post-mortem with secondary ion mass spectroscopy (SIMS), Auger electron spectroscopy (AES) and Rutherford backscattering spectroscopy (RBS) after being transported to the dedicated analysis chambers [1,2]. While it is possible to use RBS for nondestructive in-situ analysis (as in the DIONISOS setup, for example [3]), it requires the use of expensive accelerators. Additionally, the depth resolution, which is determined by the energy resolution of the detector, can only be adjusted to several tens or hundreds of nanometers, which is insufficient in some cases.

For instance, it was found in modeling the experiments on the linear simulator PR-2 [4] that formation of a thin (~10 nm) oxide layer on the aluminum target strongly increases electron emission

[3] To whom any correspondence should be addressed.

under the plasma impact. This enormously enhanced electron emission can cause different plasma-surface instabilities.

For this reason, it is vital to implement an *in-situ* analysis method for determination of the composition and thickness of ultra-thin (~5-50Å) surface layers in the course of the surface treatment by plasma or its components. In this paper, the influence of the working gases (during operation of the plasma source) on the energy spectra of the reflected and recoil ions is studied.

2. Experimental setup

Singly scattered ions or recoil ions form narrow peaks in the energy spectra and provide information about the atomic composition of the outer atomic layer [5]. Moreover, it was shown [6] that the thickness of the thin layers of light elements on the surfaces of heavy substrates can be analyzed with a good depth resolution (~0.3 nm) using scattering of the hydrogen ions with the energies in the keV range.

A schematic of the experimental setup is shown in figure 1. This setup of the modernized ion mass monochromator [7] allows us to separate ions with the mass/charge ratios of up to M/Z = 100 at the maximum accelerating voltage of 40 kV. A heated duoplasmatron ensuring a low level of impurities, with an automated power supply and gas feed is used as the ion source. When operating with helium, the current of He+ ions is two orders of magnitude higher than the total current of all other components. The relative energy spreading in the beam at 5 keV is less than 5×10–3, and its angular divergence is ~0.01 rad.

The diagnostic beam line is differentially pumped with two 70 l/s turbomolecular pumps located immediately after the duoplasmatron and before the aperture into the interaction chamber. The interaction chamber is degassed with a turbomolecular pump with the output of ~103 l/s. This feature makes it possible to use a built-in plasma/ion source based on the Penning-type discharge [8]. It can be positioned in the interaction chamber at variable distance (20-40 mm) from the target surface. This source is capable of working at relatively low pressures (~10–4 Torr) with the maximum current density of 100 μA/cm2 (Ar).

A quarter-spherical remote deflector is used as the energy spectrum analyzer for the ion scattering and recoil ion spectroscopy. The ion current is measured with a secondary electron multiplier. This detector is positioned in the forward direction at the scattering angle of θ = 16°. At the solid angle of scattered particle recording of 1.1×10–3 sr with the half-width of the scattering angle Δϑ = ±0.6°, its energy resolution is 0.8%. To keep the secondary emission multiplier below the maximum operating pressure (10–5 Torr) by increased gas puffing into the plasma source, the analyzer is equipped with an additional turbomolecular pump with the 70 l/s pumping speed. The targets, up to 40 mm in diameter, can be replaced without opening the facility by means of a lock system with a self-contained evacuation system. The target is heated with a 120 W incandescent lamp to the temperature of 750°C. The automated ion beam control makes it possible to perform experiments following predeveloped irradiation programs. Computer control is realized with the Labview and DCON Utility programs using the ADC/DAC manufactured by ICP DAS.

The residual gas composition is controlled by the Extor XT100M quadrupole mass-spectrometer.

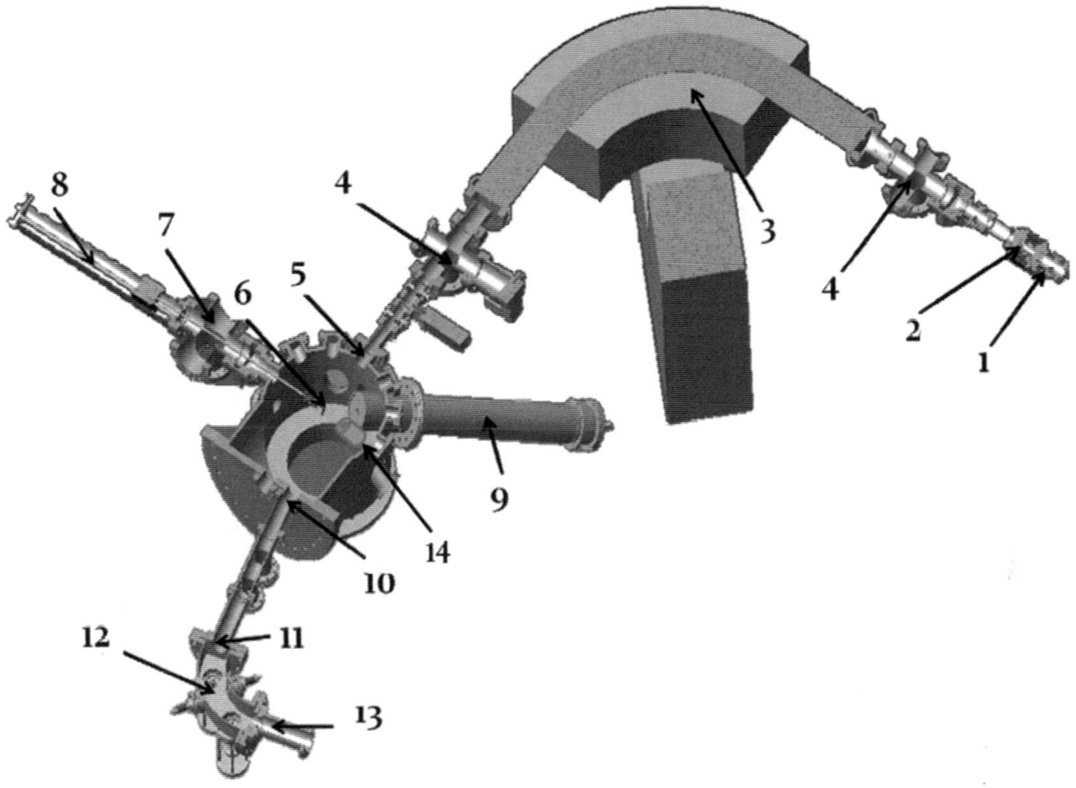

Figure 1. Schematics of the experimental setup: (1) duoplasmatron,(2) electrostatic lens, (3) separating magnet, (4) vacuum chamber for differential evacuation, (5) input diaphragm of the principal ion beam, (6) target, (7) vacuum chamber for target replacement, (8) target in–out manipulator, (9) residual gas composition controller – a quadrupole mass-spectrometer, (10) input diaphragm of the energy analyzer, (11) output diaphragm of the energy analyzer, (12) electrostatic energy analyzer, (13) secondary emission multiplier VEU-1, (14) embedded plasma/ion source based on the Penning-type discharge. Varian SH-100 forepumps and Pfeiffer TMU 071, TMU 256, and TMH 1500 turbomolecular pumps are not shown in the figure.

In this experimental setup, it was shown that when the target is irradiated with argon ions, registration of the negatively charged recoil ions makes it possible to monitor the dynamics of carbon and oxygen impurity variations on the surfaces of materials with medium and large atomic numbers. Irradiation with hydrogen ions and registration of the positively charged ions allows us to estimate the thickness of the deposited layers by the shift of the maximum of the energy spectrum and the comparison with the spectra simulated with a TRIM-like code.

3. Results and discussion

During the plasma source operation, the pressure in the vacuum chamber can increase up to 10^{-4} Torr. Prior to the experiments with the plasma source, the influence of the elevated gas pressure on the energy spectrum of the reflected ions and direct recoil ions was studied. Various gases (H_2, Ar, He, Xe) were injected into the vacuum chamber through the mass flow controller of the plasma source. The signal intensity of the H^+ ions scattered from the surface (with the initial energy $E_0 \approx$5-10 keV) increased with increasing the vacuum pressure (see figure 2). At higher pressure, a peak corresponding to single-scattering of the primary ion beam on the gas-phase appears. This peak can be used as the additional reference point for calibration of the primary beam energy. For example, the energy spectra of the H^+ ions scattered from the Al surface in Ar and Xe gas environments are shown in figures 2 and 3. As can be seen from figure 3, the signal intensity of the scattered ions drops sharply at the pressure of ~10^{-3} Torr. The obtained spectra were normalized to the primary beam current and secondary ion-electron emission was taken into account. The secondary ion-electron emission coefficient was assumed to be the same for the diagnostic ion beam and the fast neutrals resulting from the charge exchange with the gas.

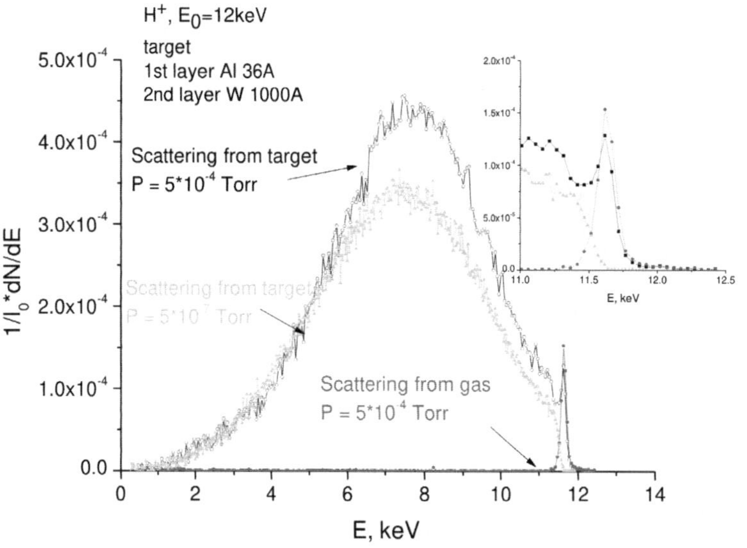

Figure 2. Ion scattering energy spectra for the 12 keV H^+ beam incident on W with the Al layer (4nm) at different Ar pressures. (I_0 is the primary ion beam current). The energy spectrum under the low pressure P=5·10^{-7} Torr is shown in green, the energy spectrum under the high pressure P=5·10^{-4} Torr in black and the energy spectrum of the H^+ ion scattered from the gas-phase without W target under the high pressure P=5·10^{-4} Torr in red.

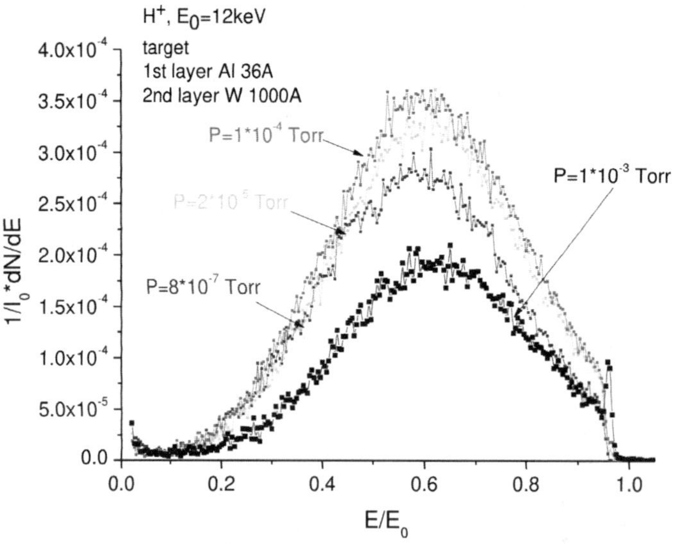

Figure 3. Ion scattering energy spectra for the 12 keV H^+ beam incident on W with the Al layer (4nm) at different Xe pressures (I_0 is the primary ion beam current).

The change of the positive charge fraction η+ during ion beam propagation through the gas-phase was estimated. The ratio of the number of the positively charged particles to the total number of particles when the beam passes through a gas target can be written as:

$$\eta_g^{\ +} = \eta_{eq}^{+} + (\eta_0^{+} - \eta_{eq}^{+}) \cdot \exp(-(\sigma_i + \sigma_n)\int_0^z n\mathrm{d}z) \qquad (1)$$

where $\eta_+^{\ 0}$ is the initial positively charged fraction of the beam reflected from the target, $\eta_{eq}^{+} = \sigma_n / (\sigma_i + \sigma_n)$ is the equilibrium value of the positive fraction at high pressures (for a "thick" target, $(\sigma_i + \sigma_n) nl >> 1$). $\sigma_n(E)$ is the ionization cross-section of the neutral atom, $\sigma_i(E)$ the cross section of the electron capture by the positive ion, z is the coordinate along the direction of beam propagation in the gas target and E the kinetic energy of the ion.

In view of the beam attenuation due to scattering, the fraction of the surviving ions that enter the analyzer can be expressed as:

$$\eta^{+} = \eta_g^{\ +} \cdot \exp(-\sigma_p \int_0^z n\mathrm{d}z) \qquad (2)$$

where $\sigma_p(E)$ is the transport cross section for the interaction of the incident ion and a neutral particle.

As shown in [9], the cross section for this type of interaction can be expressed as follows:

$$\sigma \approx 2\sqrt{2}\pi a_0^2 \sqrt{\frac{\alpha}{a_0^3} \frac{I_H}{E}} \qquad (3)$$

where a_0 =0,529·10^{-8} cm is the Bohr radius, I_H= 13.6 eV is the ionization potential of hydrogen, α is the polarization efficiency, E is the kinetic energy of the ion. The value of the polarization efficiency, or rather, the coefficient $\alpha/a_0^{\ 3}$ for basic residual gas molecules and atoms of inert gases can be found in [9].

Taking into account the length of the ion path including the interaction chamber l_c, the distance from the input aperture to the energy analyzer l_a , the pumping speed of the vacuum chamber and the analyzer, and the size of the analyzer input aperture, the "thickness" of the target gas (argon in this case) can be estimated as

$$\int_0^z n\mathrm{d}z = n_c l_c + n_a l_a \qquad (4)$$

where n_c ,n_a are the concentrations of particles in the interaction chamber and in the analyzer, respectively.

To estimate the cross sections of the electron capture and loss for hydrogen in argon, one can use the data from [10]. As shown in [11], the fractions of the neutral and charged components in the beam reflected from the target are related through the semi-empirical formula:

$$\eta_0^{\ +} = k \cdot E^{\mu} \qquad (5)$$

where k≈0.08, μ≈0.5 and E is the ion kinetic energy in keV.

If the dependence (5) is assumed, the positive charged fraction decreases by 2-3%. If we assume the values of the neutral and positively charged components of the beam reflected from the target to be $\eta_0^{\ +}$=5%, $\eta_0^{\ 0}$=95%, which are the values from another example, then the positively charged fraction is estimated to increase by 3-4%. However, as shown in figure 2, the charge fraction is increased by 7-10%. A possible explanation for this result is that a large fraction of the neutral hydrogen atoms are reflected from the surface in the excited state [12], and for such atoms the ionization cross-section is significantly higher than that for the ground-state atoms.

The signal intensity of the negative ions rebound and sputtered from the Ar $^+$-irradiated surface decreases by a factor of ~5 when the pressure in the interaction chamber increases by more than two orders of magnitude (see figure 4), as a result of capture of the electrons by the negative ions in the residual gas. At the same time, the shape of the spectrum does not change. This is in a qualitative agreement with [13,14]. However, at the elevated pressure, the signal / noise ratio for the recoil ion peaks is greater in the case of negative ion registration than it is for the positive ions.

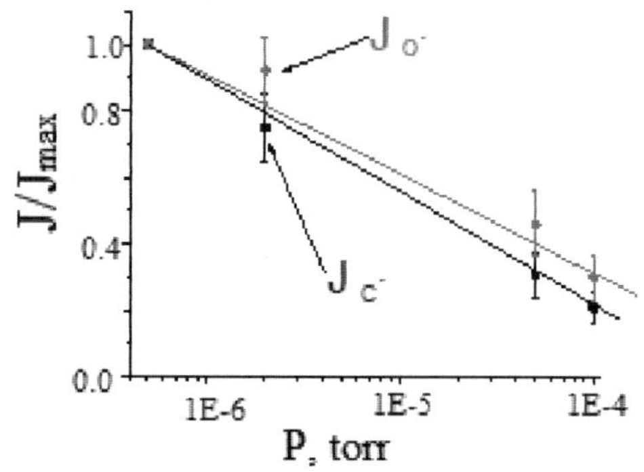

Figure 4. Signal intensity of negatively charged recoil ions of carbon and oxygen from W during irradiation by Ar^+ with energy of 7.5 keV, as function of the pressure in the interaction chamber.

4. Conclusions

Possible application of ion scattering spectroscopy at pressures up to several mTorr during plasma irradiation is described. It is found that:

- The increase of the positive fraction recorded by the energy analyzer is attributed to enhanced electron stripping from the reflected neutral atoms (especially from the excited atoms) of hydrogen in argon.
- The increase in intensity and the broadening of the reflection spectrum of light ions in the low-energy range results in a higher sensitivity for the analysis of the sample depth.

Due to neutralization of the singly reflected or recoil ions, a two orders of magnitude increase of the pressure causes reduction of the sensitivity of the elemental analysis by ion scattering by a factor of 3-5, depending on the concentration of the working gas in the interaction chamber.

References

[1] Goeble D M, Campbell G and Conn R W 1984 *J Mucl.Mater.* **121** 277
[2] Rapp J 2010 *Proc.International workshop on requirements for next generation PMI stands in fusion research* Oak Ridge USA
[3] Wright G M, Whyte D G and Lipschultz B 2009 *J Mucl.Mater.* **390** 544
[4] Gutorov K M, Vizgalov I V, Markina E A and Kurnaev V A 2010 *Bul. of the Rus. Academy of Sciences: Phys.* **74** 188
[5] Brongersma H H, Draxler M, de Ridder M and Bauer P 2007 *Surface Science Reports* **62** 63.
[6] Trifonov N.N. *et al* 2000 *Vacuum* **56** 253
[7] Mamedov N V, Ivanov D V, Kurnaev V A and Sinel'nikov D N 2012 *Bul. of the Rus. Academy of Sciences: Phys* **76** 683
[8] Mamedov N V, Kurnaev V A, Vizgalov I P and Sinelnikov D N 2013 *Nuclear physics and engineering* **4** 436 (in Russian)
[9] Raizer Yu.P. 1987 *Physics of gas discharge* (Moscow: Nauka) (in Russian)
[10] Novikov N V and Teplova Ya A 2009 Database on charge – changing cross sections in ion – atom collisions *J. Phys.: Conf. Ser.* **194** 194
[11] Kurnaev V A, Protasov Yu S and Tsvetkov I V 2008 *Introduction to beam electronics* (Moscow: MEPhI) (in Russian)
[12] Pleshivtsev N V and Bazhin A I 1998 *Physics of the ion beam impact on materials* (Moscow: Vuzovskaja Kniga) (in Russian)
[13] Jaemo Im *et al* 1996 *Nucl. Instr. And Meth. in Phys. Res. B* **118** 772
[14] Yuping Lin *et al* 1994 *Thin Solid Films* **253** 247

Journal of Physics: Conference Series **666** (2016) 012028 doi:10.1088/1742-6596/666/1/012028

Evaluation of plasma pressure of high current low inductance vacuum spark on cathode surface

S A Sarantsev[1]
National Research Nuclear University MEPhI, Kashirskoe shosse 31, Moscow, 115409, Russian Federation

E-Mail: sarsa1305@mail.ru

Abstract. This paper presents evaluation of the plasma pressure in a high current, low inductance vacuum spark on the cathode surface (the electrode material is steel). Calculations are provided for the first half period of the discharge, wherein the cathode surface is subjected to the most severe impacts (micropinches are created resulting in high-energy plasma beams). The evaluations were made using the experimental data obtained on the Pion device. The data of electrical measurements of the discharge current, the average plasma flow values obtained with the multi-grid probe and the data from a cathode macrostructure study were used. The results are given for different values of the discharge current.

1. Introduction

Pulsed electrical discharges, where the micropinch mode occurs, have been the subject of intense research for quite a long time due to a wide range of applied science missions ranging from the controlled thermonuclear fusion (CTF) to the development of selective shortwave radiation sources: ultra-violet (UV), vacuum ultra-violet (VUV) and soft X-ray (SXR). The micropinch phenomenon is quite common for all types of Z-pinches. The low inductance vacuum spark (LIVS) [1-5] belongs to this class of discharges.

Many practical applications of such a plasma discharge require data on the plasma decay mechanisms (in the inter-electrode spacing and beyond) and the parameters of the plasma and corpuscular flows from the high intensity, low inductance vacuum spark discharge area. Among these applications is an effective source of the VUV and SXR radiation for the X-ray lithography and X-ray microscopy. Plasma fluxes from the discharge area can be characterized by the value of the gas-kinetic pressure. The data on the pressure in the plasma fluxes from the micropinch discharge allow to estimate the thermodynamic parameters and the plasma flows.

This paper presents the study of the plasma pressure on the cathode surface in the high intensity, low inductance vacuum spark discharge area. For the evaluation, the experimental data obtained on the Pion device [6] of the Department of Plasma Physics of MEPhI are used.

2. Evaluation of plasma pressure on the cathode surface

The electrical characteristics of the discharge are the oscillatory circuit characteristics, where the energy is transformed from the electric to the magnetic form and back. With time, the energy decays

[1] To whom any correspondence should be addressed.

Journal of Physics: Conference Series **666** (2016) 012028 doi:10.1088/1742-6596/666/1/012028

because of the Ohmic losses, radiation losses and losses associated with the particle escape from the inter-electrode spacing. Figure 1 shows the energy evolution in the high intensity, low inductance vacuum spark discharge for different values of the charging voltage. The stored energy decreases by more than a factor of two during the first half-period. A micropinch is formed during this time interval and hence the accelerated particle fluxes are formed. Their intense interaction with the electrode surface leads to electrode spraying and melting. Some part of the melt layer is forced by the plasma pressure out of the area of direct interaction with the plasma. This area is marked in figure 2.

One can see that outside this area, the material eruptions in the form of thin solidified metal jets are generated. In the assessment, it is assumed that the impact of the plasma pressure on the cathode is limited to the direct interaction area. The diameter of this area decreases as the battery voltage decreases (figure 3).

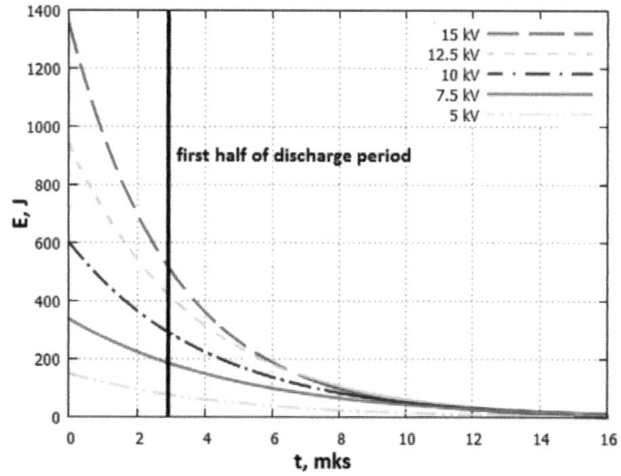

Figure 1. Decay of the discharge energy E(t) for different voltages across the inter-electrode gap.

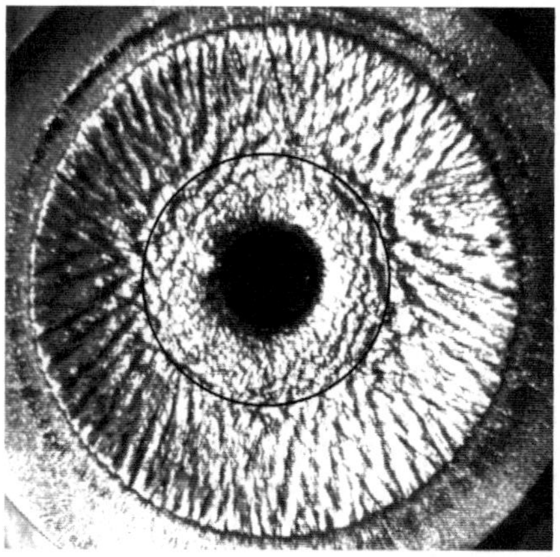

Figure 2. Macroscopic structure on the cathode surface formed after 50 discharges at a charging voltage of 15 kV (the thin circular line marks the area of the direct interaction with the plasma).

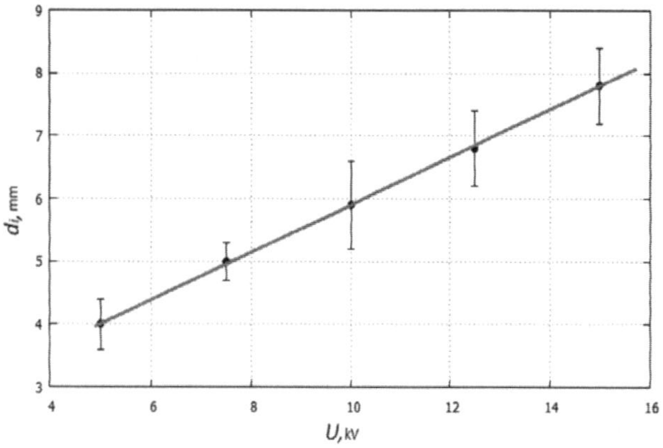

Figure 3. The diameter of the cathode area subjected to the direct interaction with the plasma flows depending on the voltage across the inter-electrode gap.

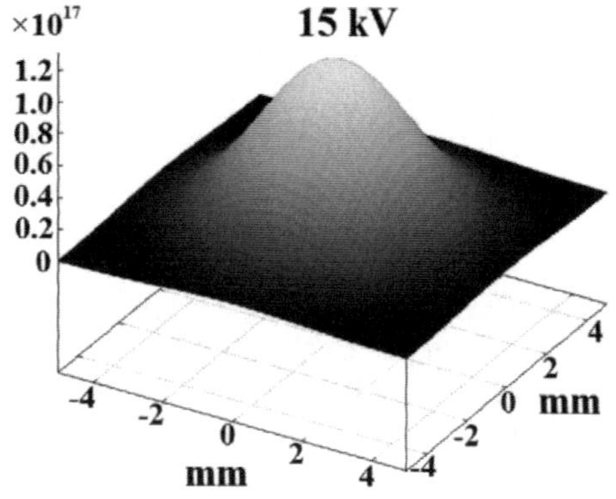

Figure 4. Example of distribution of the flow of ions bombarding the cathode over the electrode surface (Gaussian distribution) for the voltage of 15 kV.

Journal of Physics: Conference Series **666** (2016) 012028 doi:10.1088/1742-6596/666/1/012028

The distribution of the flow of ions bombarding the cathode is not uniform over the electrode surface. The highest density of the incident plasma flow is on the central area of the cathode. For the evaluation, we can take the Gaussian distribution (figure 4) for the distribution of the total number of particles falling on the cathode *D(x,y)*:

$$D(x,y) = D_0 \cdot e^{\frac{-(x-x_0)^2}{2\sigma^2}} \cdot e^{\frac{-(y-y_0)^2}{2\sigma^2}} \quad (1)$$

Here D_0 is the peak value (the number of ions arriving at the centre of the cathode), x_0 and y_0 the cathode centre coordinates and σ the parameter of the normal distribution.

Distribution parameters x_0, y_0, D_0 and σ are determined as follows. After 50 discharges, the working part of the electrode is identified visually, and x_0, y_0 are taken as one half of its diameter d_i, table 2.

The σ value is determined according to the rule of 'two sigma', so that all the values of the variables *x* and *y* with a probability of 95% fall within the area of the diameter d_i. The peak value of distribution D_0 is set to make the total number of particles coming to the cathode N_Σ equal to the integral of *D(x,y)* over the area of the diameter d_i.

Table 1. Parameters of the distribution functions of ions bombarding cathode.

U (kV)	x_0, y_0 (mm)	d_i (mm)	σ (mm)	*K*	$D_0, (\frac{\text{particles}}{\text{mm}^2})$
5	2.00	4.0±0.4	1.00	2.39	$(1.3\pm0.2)\cdot10^{17}$
7.5	2.50	5.0±0.3	1.25	2.99	$(1.2\pm0.2)\cdot10^{17}$
10	2.95	5.9±0.7	1.48	3.54	$(1.0\pm0.3)\cdot10^{17}$
12.5	3.40	6.8±0.6	1.70	4.07	$(1.1\pm0.4)\cdot10^{17}$
15	3.90	7.8±0.6	1.95	4.67	$(1.0\pm0.4)\cdot10^{17}$

The flux of ions bombarding the cathode may be represented through the current flowing through interelectrode gap, in assumption, that all ions are once ionized. The density of the flux of ions may be written as:

$$j(x,y,t) = j_0(t) \cdot e^{\frac{-(x-x_0)^2}{2\sigma^2}} \cdot e^{\frac{-(y-y_0)^2}{2\sigma^2}} \quad (2)$$

$j_0(t)$ value may be calculated through particle flux, which is the integral over interaction domain (region with di diameter) of the discharge current:

$$J(t) = \int^{d_i}\int^{d_i} j(x,y,t)\mathrm{d}x\mathrm{d}y = \int^{d_i} j_0(t) \cdot e^{\frac{-(x-x_0)^2}{2\sigma^2}} \mathrm{d}x \cdot \int^{d_i} e^{\frac{-(y-y_0)^2}{2\sigma^2}} \mathrm{d}y$$

Then $j_0(t)$ is:

$$j_0(t) = \frac{J(t)}{qK^2}, \quad (3)$$

Where substitution is made:

$$K = \int_0^{d_i} e^{\frac{-(x-x_0)^2}{2\sigma^2}} dx = \int_0^{d_i} e^{\frac{-(y-y_0)^2}{2\sigma^2}} dy. \quad (4)$$

Then:

$$j(x,y,t)=\frac{J_0\cdot\sin(\omega t)}{qK^2}\cdot e^{\frac{-(x-x_0)^2}{2\sigma^2}}\cdot e^{\frac{-(y-y_0)^2}{2\sigma^2}}=10^6\,\frac{J_0\cdot\sin(\omega t)}{qK^2}\cdot e^{\frac{-(x-x_0)^2}{2\sigma^2}}\cdot e^{\frac{-(y-y_0)^2}{2\sigma^2}}\left[\frac{A}{coulomb\cdot m^2}\right] \quad (5)$$

where ω is the frequency of the discharge current oscillation, J_0 – is its amplitude, q – is the ion charge (presumed single).

By expressing the pressure P as the flux density times the momentum p carried by a single particle to the cathode surface, it can be written as follows:

$$P(x,y,t)=j(t)\cdot p=j(t)\cdot\sqrt{2m\langle E\rangle}=10^6\cdot\frac{J_0\cdot\sqrt{2m\langle E\rangle}}{qK^2}\cdot e^{\frac{-(x-x_0)^2}{2\sigma^2}}\cdot e^{\frac{-(y-y_0)^2}{2\sigma^2}}\cdot sin(\omega t),$$
$$P(r,t)=10^6\cdot\frac{J_0\cdot\sqrt{2m\langle E\rangle}}{qK^2}\cdot e^{\frac{-r^2}{2\sigma^2}}\cdot sin(\omega t)[Pa] \quad (6)$$

The ion energy component in the axial discharge direction measured with a multi-grid probe is used for the calculation of the average ion energy. The dependence of the average ion energy on the charging voltage is shown in figure 5. Using these data, the maximum pressure in the centre of the cathode (figure 6) is calculated. As an example, figure 7 shows the pressure in the plasma flow as a function of time and distance from the discharge axis for the charging voltage of 15 kV.

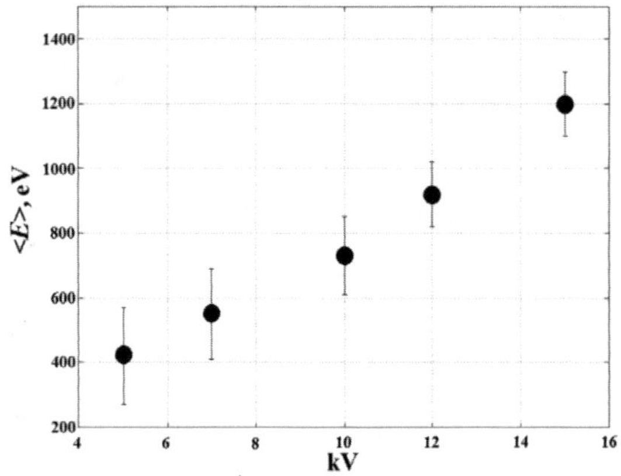

Figure 5. The value of the average energy of ions depending on the voltage at the inter-electrode gap.

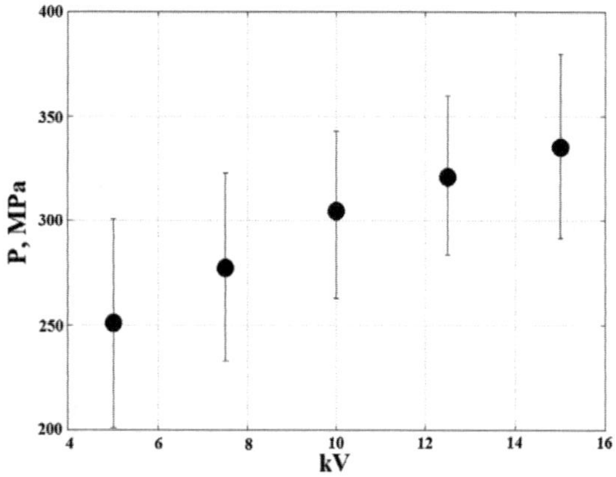

Figure 6. Maximum values of the plasma flow pressure on the cathode surface depending on the voltage at the inter-electrode gap.

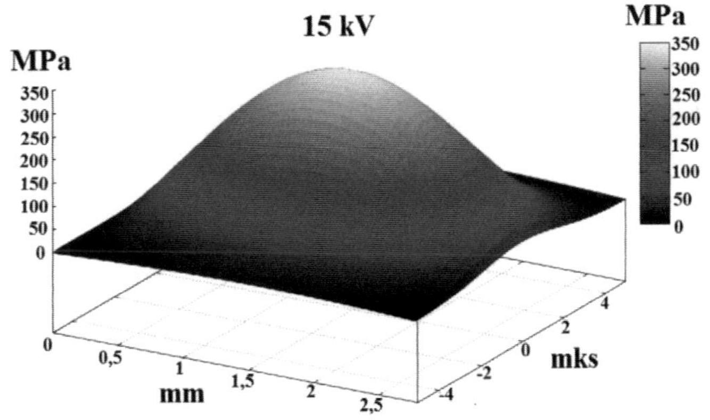

Figure 7. Pressure as function of time and distance from the axis in the first half period of the discharge with the charging voltage of 15 kV.

3. Conclusions

The evaluation of the pressure of the plasma flows shows that the cathode surface is subjected to a considerable load that drives the molten cathode material outside the area of direct interaction with the plasma flow.

In [1], the radial component of the plasma pressure at the edge of the high intensity, low inductance vacuum spark was measured to be of several MPa. The evaluations presented here indicate that in the central part of the cathode exposed to the plasma flows, the plasma pressure exceeds these values by up to two orders of magnitude.

References

[1] Kuznetsov A P, Byalkovskii O A, Gubskii K L, Kozin G I, Protsenko E D, Dodulad E I and Savjolov A S 2014 *Plasma Physics Reports* **40** 290-297

[2] Dodulad E I, Kostyushin V A and Savjolov S A S 2015 *Physics Procedia* **71** 155-159

[3] Krasov V I, Paperny V L , Korobkin Yu V, Romanov I V, Rupasov A A and Shikanov A S 2007 *Technical physics letters* **33** 941

[4] Bashutin O A, Alkhimova M A, Vovchenko E D, Dodulad E I, Savelov A S and Sarantsev S A 2013 *Plasma Physics Reports* **39** 900-909

[5] Astrachantsev N V, Krasov V I and Paperny V L 1995 *Journal of physics D: Applied physics*, **8** 2514

[6] Sarantsev S A, Dvoyeglazov Ya M and Raevskiy I F 2015 *Physics Procedia* **71** 133-137

Journal of Physics: Conference Series **666** (2016) 012029 doi:10.1088/1742-6596/666/1/012029

Energy and matter flows in a plasma focus discharge

V V Vikhrev and S V Suslin[1]
National Research Centre "Kurchatov Institute", Kurchatov square, 1, Moscow, 123182, Russian Federation

E-mail: SuslinSV@yandex.ru

Abstract. The Plasma Focus is a type of z-pinch that is widely used for both basic research and applied tasks, e.g., as materials modification or research on intense plasma flows. Although the basic mechanisms of z-pinch compression are well-known, many of the processes that occur in the plasma focus have received less attention. This article is devoted to the study of plasma jets and some of its consequences in plasma focus discharges.

1. Introduction

Compared to similar types of pulsed discharges, the Plasma Focus (PF) possesses a number of advantages for the generation of neutrons and x-rays over other pulsed discharges, including the relative simplicity in the manufacture of the discharge chamber and the high efficiency of neutron and X-ray generation. Furthermore, plasma focus can generate a jet of plasma jet moving at a speed up to 10^7 cm/s. The latter feature can be used for laboratory simulation of astrophysical jets - narrow streams of plasma, up to several megaparsecs [1].

Initially, the plasma focus was a very promising line of research in nuclear fusion. Its neutron yield Y increased very rapidly with peak current I, as $Y{\sim}I^4$; the tentative conclusion was that the the neutron yield, and therefore the thermonuclear energy produced, is the square of the input energy. However, it soon became clear that the neutron yield saturates as the energy input approaches about 1 MJ [2].

Despite its long history, the processes underlying plasma focus formation are rather poorly studied. This is in part due to the short time scale of the relevant processes – at most a few microseconds - which are difficult to diagnose. As a result, optimization of the plasma focus for the different applications of interest still requires much further study.

The following research describes attempts to investigate the dynamics of deuterium plasma and energy flows in the plasma focus with numerical simulation, in the two-dimensional single-fluid MHD approximation.

2. Simulation results

The simulation shows two types of plasma flows: a jet extending along the axis of the chamber after the formation of the plasma focus; and the flow of rarefied plasma along the anode, which is observed during the whole discharge period.

Figure 1 shows the plasma parameters 1.44 microseconds after the start of discharge or 0.31 microseconds after the formation of the plasma focus proper. The plasma density is shown in the top of the upper plot (figure 1a). The axis of the machine contains a narrow plasma filament that is compressed by a magnetic field shown in the bottom part of the upper plot (figure 1b) This plasma

[1] To whom any correspondence should be addressed

feature is moving at a speed of about 10^7 cm/s (in the top of the lower plot (figure 1c)). Apparently, this speed is maintained by the pressure gradient of the magnetic field, seen in the bottom of the lower plot (figure 1b). The jet in this simulation is similar to the axial jet that is observed experimentally in a number of installations, such as the PF-4 ("Tulip", LPI RAS) [3].

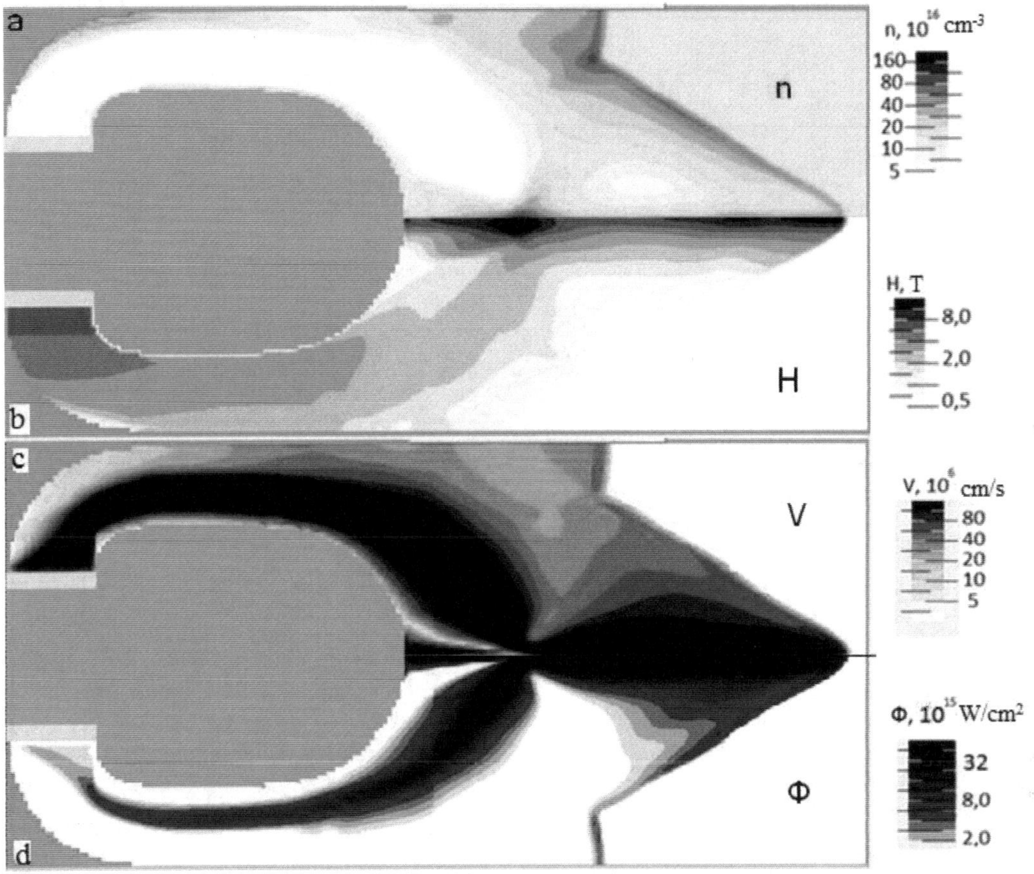

Figure 1. Distribution of plasma density (a), magnetic field strength (b), plasma velocity modulus (c), total energy flow density (d); t=1,44 microseconds.

In addition to the plasma jet on axis, a high velocity but tenuous plasma is observed to flow along the anode, toward the insulator to the left in the figure. The flow rate reaches 10^8 cm/s (figure 2), which exceeds the Alfvén velocity that is estimated to be up to $5 \cdot 10^7$ cm/s in this region. Such high velocities cannot be explained by the acceleration of the plasma by a magnetic piston, but can occur with the flow in a narrow channel. In [4] the conditions are described, under which the rate of flow of magnetized plasma in a narrow channel can exceed the speed of magnetosonic wave.

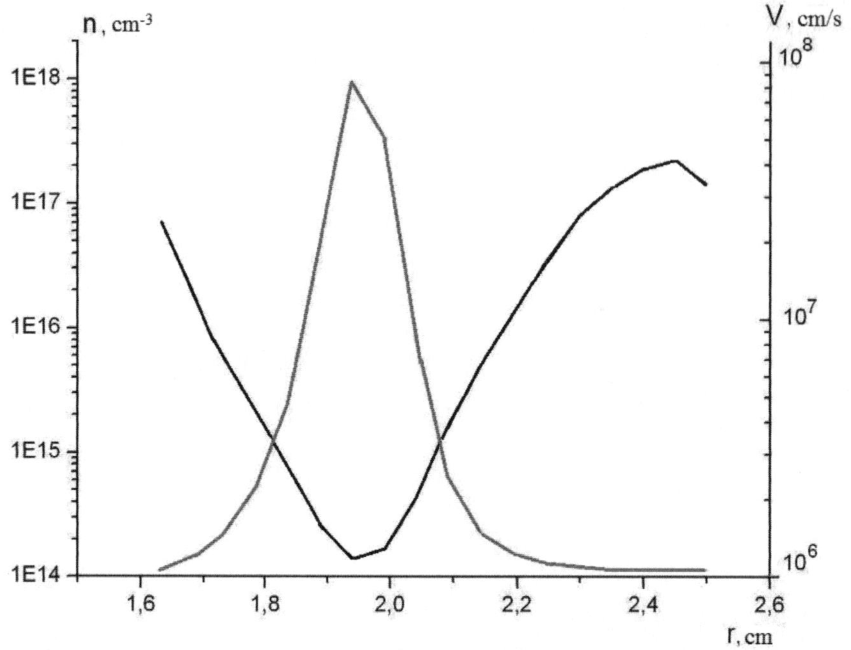

Figure 2. Density (black graph) and velocity (red graph) distribution in cross section z=2,5cm; at t=1.44 microseconds.

Unfortunately, the diagnosis of the pre-anode flow is difficult - low plasma density does not allow the use of optical methods, and the movement of dense current-carrying plasma sheath (CPS) at the initial stage of the discharge makes it difficult to use probes. However, this flow has a significant impact on the dynamics of the discharge, since it transports energy, both magnetic and kinetic, from the insulator to the current-carrying plasma sheath. A similar flow is also observed before the pinch formation. Thus, in figure 1d an uneven energy flow can be seen propagating along the anode. The energy flow is concentrated in a narrow channel, which affects the shape of the current-carrying plasma sheath before the pinch.

The reason for the flow collimation can be understood from (figure 2) - flow rate is limited by the inertia of plasma, and consequently, it reaches maximum in the minimum density area. This leads to a more rapid plasma sweeping and to further density reduction so that finally a narrow channel of low-density plasma is formed, in which the flow velocity reaches maximum.

References

[1] Marscher A P and Jorstad S G 2011 The megaparsec-scale x-ray jet of the BL LAC object OJ287 *ApJ* **729** 26

[2] Nikulin V Y and Polukhin S N 2007 On saturation of the plasma focus neutron yield in megajoule range *Plasma Physic* **33** 304

[3] Baronova E O et al 2012 *Plasma Physics Reports* **38** 751

[4] Morozov A I 2006 *Introduction to plasmodynamics* (Moscow: FIZMATLIT) p 125

Reconstruction of particle energy spectrum for angular dependent distribution at the entrance of electrostatic spectrometer

V A Urusov[1]
National Research Nuclear University MEPhI, Kashirskoe shosse 31, Moscow, 115409, Russian Federation

E-mail: urvik@mail.ru

Abstract. The algorithm of reconstruction of particle energy spectrum from electrostatic spectrometer data under condition of angular dependence of the energy distribution of the charged particles flow at the spectrometer entrance. It is shown that, assuming a linear relationship between the potentials, a standard method of recovering the spectrum by means of dividing the recorded signal by the scanning voltage gives the correct result in the first approximation.

If the energy distribution at the electrostatic spectrometer entrance does not depend on the angular distribution, the reconstruction of spectrum using the standard procedure of dividing the recorded signal by the scanning voltage gives the correct result in the first approximation [1, 2]. At the entrance of electrostatic dispersion spectrometers, particles have different angular distribution at different energies. These spectrometers therefore have different instrument functions, and the procedure of spectrum reconstruction using these spectrometers has not been justified yet. Meanwhile, for a large number of such spectrometers used, for example, in the mass spectrometry, the condition of the energy distribution independence from angular distribution fails.

To justify the recovery procedure of the spectrum we will use the results from Ref. [1, 2]. The signal at the spectrometer outlet is related to the particles energy and angular distribution function by the equation

$$I = I_0 \int_0^{\infty} \int_{S_0} \int_{S_1} \delta(E) f(\eta, \xi, \alpha, \beta, E) J(\alpha, \beta, \eta_1, \xi_1) \sin(\alpha) \mathrm{d}S_1 \mathrm{d}S_0 \mathrm{d}E, \tag{1}$$

where α and β are the incidence angles of particles at the inlet diaphragm:

$$\alpha = \alpha\left(\eta, \xi, \eta_1, \xi_1, \frac{qU_1}{E}, ..., \frac{qU_n}{E}\right), \beta = \beta\left(\eta, \xi, \eta_1, \xi_1, \frac{qU_1}{E}, ..., \frac{qU_n}{E}\right), \tag{2}$$

$$J(\alpha, \beta, \eta, \xi) = \frac{\partial(\alpha, \beta)}{\partial(\eta_1, \xi_1)}, \tag{3}$$

[1] To whom any correspondence should be addressed

where J is the Jacobian of transformation; $f(\eta,\xi,\alpha,\beta,E)$, the particle coordinates, angles and energy distribution function, normalized to unity, at the inlet aperture diaphragm; E, the particle energy; $\delta(E)$, the efficiency of particle registration by the detector at the outlet; U_i are the electrode potentials relative to the zero potential of the inlet aperture; (η_1,ξ_1), the coordinates in the outlet diaphragm plane. The integral is taken over the areas of inlet and outlet diaphragm slits.

The distribution function can be expressed as

$$f(\eta,\xi,\alpha,\beta,E)= f_1(\eta,\xi,\alpha,\beta,E)f_2(E), \tag{4}$$

where $f_2(E)$ is the particle energy distribution function, $f_1(\eta,\xi,\alpha,\beta,E)$ is the particle angular and cross-section distribution function in the inlet diaphragm plane.

Note that the angular and cross-section distribution function depends on the energy, and therefore the instrument functions for different energies will be different. Thus, if the instrument function is unknown beforehand, seeking of an exact solution is senseless. We can find an approximate solution under condition that the relative width of the instrument function is much less than unity and assuming a linear relationship between the potentials:

$$\frac{U_2}{U_1} = \lambda_2,\ldots,\frac{U_n}{U_1} = \lambda_n , \tag{5}$$

where $\lambda_2,\ldots,\lambda_n$ are the constants.

From (1) it follows:

$$I \approx I_0\delta(W)f(W)\int_0^\infty\int_{S_0}\int_{S_1} f_1(\eta,\xi,\alpha,\beta,W)J(\alpha,\beta,\eta_1\xi_1)\sin\alpha dS_1 dS_0 dE , \tag{6}$$

where W is the tuning energy of spectrometer,

$$W = qkU_1 , \tag{7}$$

$$\alpha = \alpha\left(\eta,\xi,\eta_1,\xi_1,\frac{qU_1}{E},\lambda_2\ldots,\lambda_n\right),\quad \beta = \beta\left(\eta,\xi,\eta_1,\xi_1,\frac{qU_1}{E},\lambda_2,\ldots,\lambda_n\right). \tag{8}$$

We will show that the integral in the right-hand side of (6) in the first approximation can be represented in the form

$$\int_0^\infty\int_{S_0}\int_{S_1} f_1(\eta,\xi,\alpha,\beta,W)J(\alpha,\beta,\eta_1\xi_1)\sin\alpha dS_1 dS_0 dE = CU_1 , \tag{9}$$

where C is a constant, independent from the input angular distribution.

Let us introduce the distribution function of the beam with energy E in the plane of the outlet aperture $\psi(\eta_1,\xi_1,qU_1/E,W)$ (see figure 1).

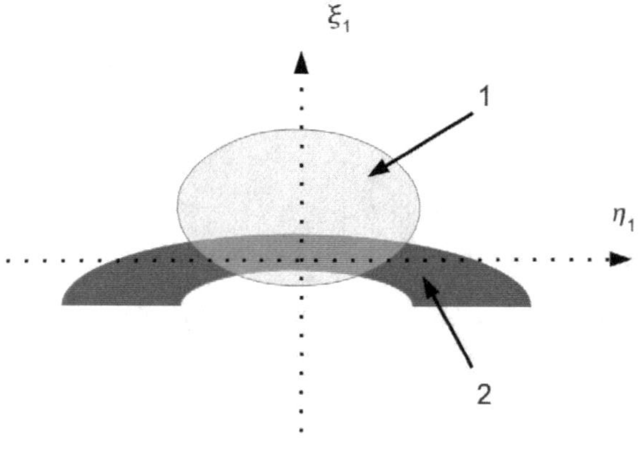

Figure 1. The distribution of particle flux in the plane of the outlet aperture: the distribution function of the particle flux (1), the slit in the outlet aperture (2).

$$\psi\left(\eta_1,\xi_1,\frac{qU_1}{E},W\right)=\int\limits_{S_0} f_1\left(\eta,\xi,\alpha,\beta,W\right)J\left(\alpha,\beta,\eta_1,\xi_1\right)\sin\alpha\,\mathrm{d}S_0\,. \tag{10}$$

Function $\psi\left(\eta_1,\xi_1,\frac{qU_1}{E},W\right)$ satisfies the condition of normalization:

$$\int \psi\left(\eta_1,\xi_1,\frac{qU_1}{E},W\right)\mathrm{d}S_1=1\,, \tag{11}$$

where the integral is taken over the entire area of the outlet aperture.

The integral of this function over the output diaphragm slit surface gives an instrument function:

$$\int\limits_{S_1}\psi\left(\eta_1,\xi_1,\frac{qU_1}{E},W\right)\mathrm{d}S_1=A\left(\frac{qU}{E},W\right). \tag{12}$$

We will assume that for small change of the parameter $z=E/qU_1$ the form of the function $\psi(\eta_1,\xi_1,qU_1/E,W)$ is not changed in the first approximation, and the shift of the function takes place along the ξ_1 coordinate:

$$\psi\left(\eta_1,\xi_1+\xi_{11},\frac{1}{z+z_1},W\right)=\psi\left(\eta_1,\xi_1,\frac{1}{z},W\right). \tag{13}$$

In fact, this condition is similar to the one formulated in the monograph [3] with, however, one significant difference: in [3] this parameter is the tuning energy *W*, whereas in our case this is the ratio of the scanning voltage to the energy of the particles or, with account of (7), the ratio of the tuning energy to the particle energy.

The integral (9) becomes

$$\int\limits_0^\infty\iint\limits_{S_1}\psi\left(\eta_1,\xi_1,\frac{qU_1}{E},W\right)\mathrm{d}\eta_1\mathrm{d}\xi_1\mathrm{d}E=qU_1\iint\limits_{S_1}\int\limits_0^\infty\psi\left(\eta_1,\xi_1,\frac{1}{z},W\right)\frac{\mathrm{d}z}{\mathrm{d}\xi_{11}}\mathrm{d}\xi_{11}\mathrm{d}\eta_1\mathrm{d}\xi_1\,. \tag{14}$$

If the relative width of the instrument function is much less than unity, the parameter z_1 and the shift ξ_{11} in the linear approximation are related by the equation

$$\frac{\mathrm{d}z_1}{\mathrm{d}\xi_{11}}\approx const=C_1\,. \tag{15}$$

The integral of the function ψ taken over ξ_{11} and η_1 equals unity. The remaining integral represents the average width of the outlet slit *D*. Consequently, the integral in the expression (9) in the first approximation is given by the equation

$$\int\limits_0^\infty\int\limits_{S_0}\int\limits_{S_1} f_1\left(\eta,\xi,\alpha,\beta,W\right)J\left(\alpha,\beta,\eta_1\xi_1\right)\sin\alpha\mathrm{d}S_1\mathrm{d}S_0\mathrm{d}E=qU_1C_1D\,. \tag{16}$$

Thus, in the first approximation the integral does not depend on the angular distribution of particles at the spectrometer entrance. Our analysis given by (9) shows that for electrostatic dispersion spectrometers with linearly related potentials, dividing the recorded signal by a scanning voltage while reconstructing the energy spectrum gives in the first approximation the correct result. It is valid even in the case when particles with different energies have different angular distribution at the analyzer inlet. We should note that the obtained results are also important for spectrometers with preretardation. The formula for the spectrum reconstruction for such spectrometers was obtained in [4, 5], where an expression similar to (9) was assumed. Namely, it was assumed that at a fixed potential bias in the dispersion part of the spectrometer, the integral over the instrument function of the dispersion part of the spectrometer will be the same, if after passing the retardation system the angular distributions of particles are different at different initial energies.

References

[1] Kurnaev V A and Urusov V A 1995 Instrument functions of electrostatic and magnetic analyzers and processing of experimental results *Pre-print* (Moscow: MEPhI) **018-95** 36
[2] Kurnaev V A and Urusov V A 1997 *Tech. Phys.* **42** 663
[3] Afanasyev V P and Yavor S Ya 1975 *Sov. Tech. Phys. Letters* **1** 227
[4] Kurnaev V A and Urusov V A 2008 *Materials of the VI Russian Seminar Modern diagnostics of plasma and their application for substances and environment monitorin*g (Moscow: MEPhI) p 127
[5] Kurnaev V A and Urusov V A 2010 *Technical Physics Letters* **36** 454

Properties of corona discharge plasma near metal surface

M Lavrinenko[1], E Biktashev[1] and D Kirko[2,3]
[1] Moscow State University of Instrument Engineering and Computer Science, Stromynka 20, Moscow, 107996, Russian Federation
[2] National Research Nuclear University MEPhI (MEPhI), Kashirskoe shosse 31, Moscow, 115409, Russian Federation

E-mail: kirko@plasma.mephi.ru

Abstract. Properties of corona discharge near metallic surface were researched. Electrical oscillations in discharge plasma of 1 kHz – 100 MHz rate were registered. Spectrum of electrical oscillations in this range was obtained. Possible plasma waves for observed electronic oscillations explanation are discussed.

1. Introduction

Nowadays a variety of corona discharges is used for treatment and modification of non-organic materials surfaces [1]. High-frequency oscillations within a wide frequency range [2, 3] is one of peculiar features of corona discharges. A typical (characteristic) electrode system normally has a sharpened electrode and a flat electrode [1, 4]. In this work we studied the properties of a corona discharge near a metal surface.

2. Experimental setup

Figure 1 shows the scheme of the experimental setup. We used a conic electrode (1) made of steel (3-4 mm in diameter, cone angle $\alpha \approx 55°$) and a cylindrical electrode (2) made of aluminum (25-35 mm in diameter, 20-25 mm thick) located at a distance d=0.5-45 mm from each other. High-voltage power supply (4) provided 0.5-25 kV at 2 mA. The conic electrode had a positive potential relative to the cylindrical electrode. A static kilovoltmeter and microampermeter were used to measure the voltage and current, respectively. The glow of corona discharge had a weak blue color.

During the experiments, two main modes were discovered. The first mode took place at d=0.5-5 mm within U=1.5-3.5 kV voltage range. The discharge current was of I=5-45 μA. The dependence of the voltage U on the distance d is nearly linear and follows the formula: $U \approx 1.1 + 0{,}76d$ (kV) (d is in cm). In this mode the corona discharge was stable. In the second mode the voltage rose up to U=4,5-25 kV for d=5-45 mm and I=60-140 μA. The corona discharge had a branched structure with a large number of narrow channels and was unstable.

[3] To whom any correspondence should be addressed.

Journal of Physics: Conference Series **666** (2016) 012031 doi:10.1088/1742-6596/666/1/012031

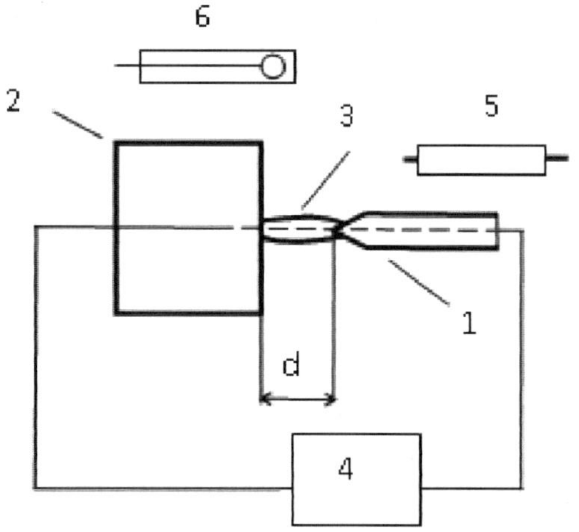

Figure 1. Experimental setup: 1 – anode, 2 – cathode, 3 – corona discharge, 4 – power supply, 5 – electrostatic probe, 6 – magnetic probe.

During the experiment, the electric oscillations were studied as well. For this purpose we used electric (5) and magnetic (6) probes connected to the Tektronix TDS 2024B oscilloscope (figure 1). Electric probes were made of metal rods (3-7 cm long and 2-4 mm in diameter) while magnetic probes were made as coils (2-4 mm in diameter, 80-160 turns, wire 0.1 mm thick). Electric and magnetic probes were 0.2-1.2 m away from the discharge. When oscillations began, the discharge current became to look like a series of separate pulses of ν=1-15 kHz frequency. Figure 2(a) shows a typical form of single pulse. A pulse had a leading edge of 15-35 μs and the falling edge of 150-200 μs. On the pulse droop an oscillations with frequency of ν=0.1-100 MHz appeared. The obtained oscillograms of pulses were processed with the Origin software suit. The most characteristic frequencies were: 4.1±0,2 kHz, 35±2 kHz, 102±5 kHz, 231±12 kHz, 2,4±0,1 MHz, 27±1 MHz, 95±5 MHz. The spectrum of corona discharge electric oscillations is presented in figure 2(b).

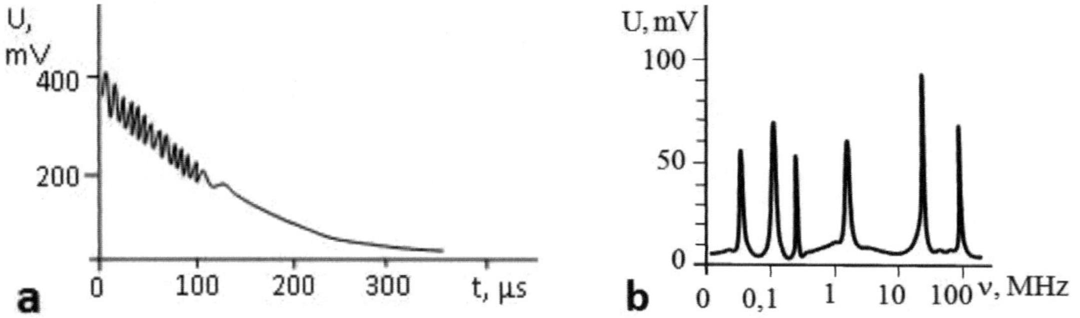

Figure 2. Electric oscillations of corona discharge: a) time dependence of the electric current, b) frequency spectrum of current's oscillations.

3. Discussion

Let us consider plasma processes in the corona discharge. For the values of plasma concentration in corona discharge n_e=10^4-10^{10} cm^{-3} the values of plasma frequency $\omega_p = (4\pi n_e e^2 / m_e)^{1/2}$ lie in the range $\omega_p = 5,5\times10^6 - 5,5\times10^9$ s^{-1}. Electron concentrations n_e=10^4-10^7 cm^{-3} in this type of corona discharge were registered in [2, 4]. The higher values, n_e=10^6-10^7 cm^{-3}, were found in the 0.3-0.5 mm wide layer around the sharpened part of the conic anode where the generation of electron avalanches is possible. Due to this fact, oscillations of the discharge within ν=1-100 MHz range can be associated with the excitation of high-frequency plasma waves in the corona discharge [5].

The materials with new reflective properties within optical and high-frequency ranges of electromagnetic irradiation are investigated in [6]. These properties are related to the phenomena taking place in a thin micron layer on the surface of a solid. According to this monograph, there is a chance that plasmon waves may exist in a metal at the frequencies lower than plasma frequency. These waves propagate in a surface layer of 10-80 μm thickness. That is why the excitation of plasmon waves with frequencies within ν=1 kHz-150 MHz range is possible in the thin surface layer of anode metal in presence of corona discharge. Thus, the electromagnetic waves with the same frequency ranges can exist both in the anode plasma layer of corona discharge and the anode surface layer. Generation of waves in plasma can be a result of solitary electron avalanches. These waves may also exist in a thin layer of an anode metal.

While studying the corona discharge near the surface of the metal we observed the oscillatory processes. The oscillation frequencies (1 kHz-150 MHz) recorded during the experiment were compared to the frequencies of plasma waves that can exist in the discharge plasma. Plasmon waves generation in a thin micron layer of the anode metal is suggested.

References

[1] Zhuravlev A A, Ivchenko V A, Bakulin I A and Strelnikov A Yu 2010 *Bar'ernaja korona postojannogo toka: processy formirovanija i primery primenenija* (Samara: Izdatel'stvo Samarskogo Gosudarstvennogo Ajerokosmicheskogo Universiteta) (in Russian)
[2] Nikitin A I, Leipunsky I O and Nikitina T F 2010 *Prikladnaya Fizika* **2** 15-22 (in Russian)
[3] Kaptsov N A 1950 *Jelektricheskie javlenija v gazah i vakuume* (M.: Gostehizdat) (in Russian)
[4] Kirko D L 2011 *Materialy IX Mezhdunarodnoj konferencii: Volnovaja jelektrodinamika provodjashhej zhidkosti. Jaroslavl'* pp 100-103 (in Russian)
[5] Kroll N A and Trivelpiece A W 1973 *Principles of plasma physics* (New York: McGrawHill)
[6] Brandt N D and Kulbachinsky V A 2007 *Kvazichasticy v fizike kondensirovannogo sostojanija* (M.: Fizmatlit) (in Russian)

Journal of Physics: Conference Series **666** (2016) 012032 doi:10.1088/1742-6596/666/1/012032

Laboratory testing of neutral particle converter device "Aries-L"

N.V. Mamedov[1, 2, 4], D.N. Sinelnikov[2], D.V. Kolodko[2], N.A. Soloviev[2], A.I. Kalinenkov[2], V.A. Kurnaev[2], O.L. Vaisberg[3], A.Y. Shestakov[3], D.A. Moiseenko[3] and R.N. Zhuravlev[3]

[1] Federal State Unitary Enterprise «All-Russia Research Institute of Automatics», Sushchevskaya street 22, Moscow, 127055, Russian Federation
[2] National Research Nuclear University MEPhI, Kashirskoe shosse 31, Moscow, 115409, Russian Federation
[3] Space Research Institute (RAS)

E-Mail: m_nikitos@mail.ru

Abstract. This paper presents the testing results of the neutral particles converter device "Aries-L" for the "Luna-Resurs" project of the Space Research Institute (RAS).

1. Introduction
Currently, within the "Luna-Resurs" project of the Space Research Institute (RAS) the "Aries-L" device is under construction [1], designed to study solar wind interaction with the surface of the Moon and for determination of its structure by means of secondary-ion mass spectrometry and spectrometry of secondary neutral atoms. The analysis of neutral atoms with electrostatic analyzers is only possible when these atoms are converted into charge fractions. Currently in space research solid-state converters are most widely used due to the ease of operation, small size and weight, as well as small distortion of the initial beam parameters [2,3]. This paper presents the testing results of the neutral particles converter device "Aries-L".

2. Description of the experiment
The experiment was conducted on the "Big monochromator of MEPhI". Neutral particles gathering occurs by the surface recharge on a golden target of positive hydrogen ions. The ion beam extracted from the source (1) is focused by the electrostatic lenses (2) and is separated according to the relation of mass to charge in the electromagnet (3) (figure 1). By varying the current in the electromagnet a necessary mass component can be isolated from the beam and directed through the aperture to the target (4) arranged in the interaction chamber. The ion current to the target is measured using a Keithley 6485 picoammeter (9). Then the reflected ions and neutral particles "light up" the input window of the converter (6) passing through the grid (5). For isolation of the ion component a positive potential up to 4000+ W could have been applied with the BNV-31 power supply. Reflected and ionized particles from the converter are focused on the input of electron multiplier-6m (7), and the number of pulses is registered with the Tektronix TDS 2004B oscillograph (8).

[4] To whom any correspondence should be addressed.

Firstly, a neutral beam was directed perpendicularly to the converter input window (i.e. with sliding pitch angle to the target $\alpha_0=8^O \leftrightarrow \beta=0$).

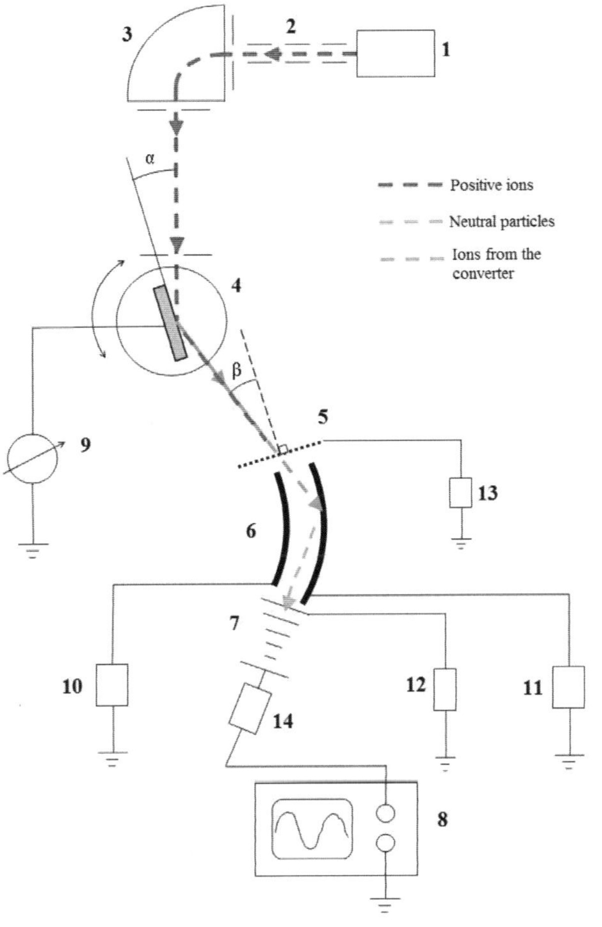

Figure 1. Experimental setup. 1 - ion source; 2 – electrostatic lenses; 3 – separating electromagnet; 4 target(Au); 5 – reflecting grid; 6 – converter; 7 – electron multiplier; 8 – oscillograph; 9 – picoammeter; 10, 11, 12, 13 – power supply; 14 - comparator.

3. Testing results

The input window of the converter was irradiated with a flux of hydrogen H atoms reflected from the target in the energy range from 200 to 1000 eV. Each value of the initial beam energy varied potentials parameters on the inner and outer plates of the converter, and thus a number of pulses has been registered on the Tektronix TDS 2004B oscillograph. Gathered results (figure 2) show that all dependencies possess a dome shape for almost all values of the initial ion beam energy. The graphs also show that the most effective operating potentials are 300 V on the inner (U_1) and 500 V on the outer (U_2) plates (except for the dependency with initial accelerating voltage of 200 eV). Also from the graphs we see that for the ratio $U_{inner} = ½U_{outer}$ the number of pulses is higher than for the ratio $U_{inner} = ⅓U_{outer}$.

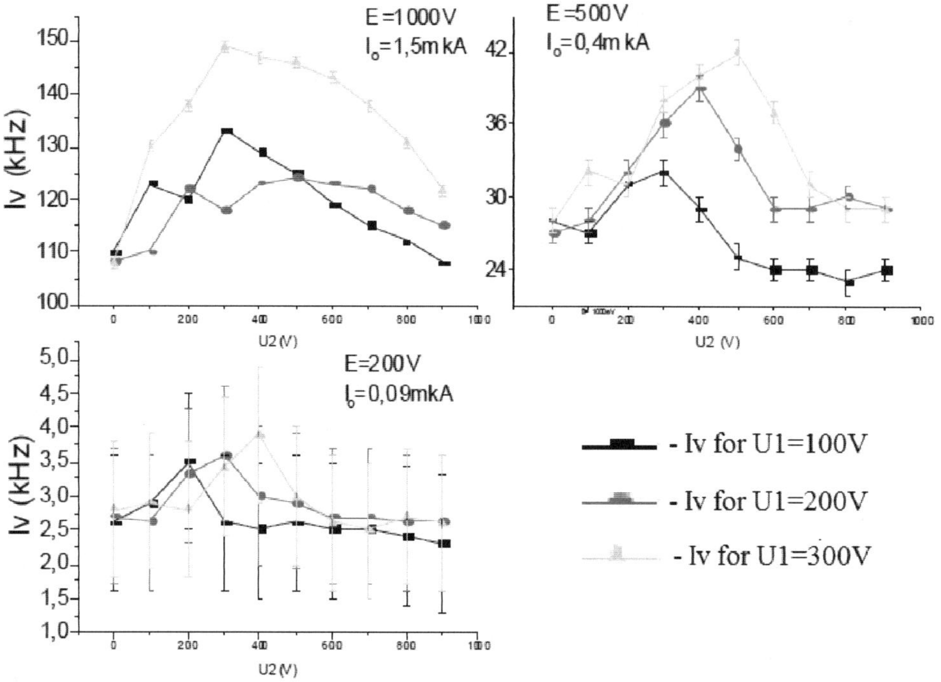

Figure 2. Working potentials variation on the converter electrodes. Pulse counting speed rates related to the voltage on the converter plates. I_v – number of pulses from the converter; I_0 – current of primary ion beam; U_1 – potential of inner plate; U_2 – potential of outer plate; E – energy of incident particle.

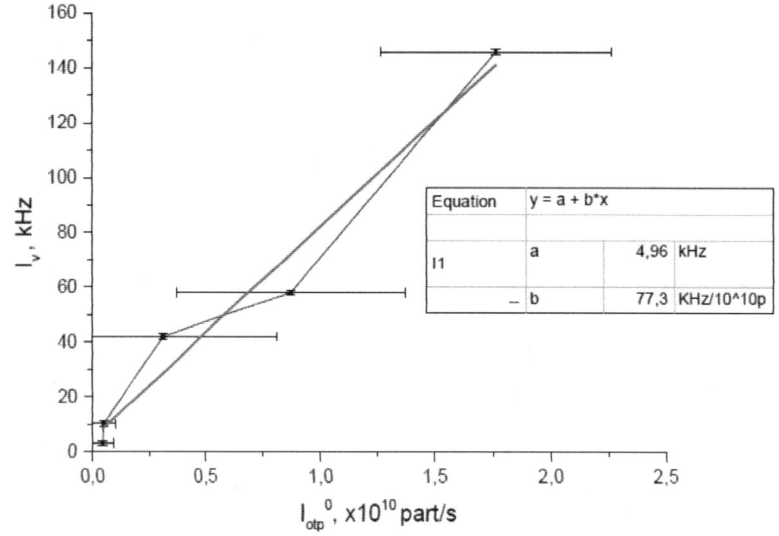

Figure 3. Correlation of neutral hydrogen atoms entering the input window of the converter to the registered counting speed of the electron multiplier.

Table 1 shows the values of the ion current to the target and the reflected neutrals flux estimate at the input of the converter depending on the accelerating voltage. A comparison of the flux of neutral particles coming into the input window of the converter with a registered counting speed of the electron multiplier is shown in Figure 3. As we can see, this dependence is linear and at the first approximation we find that at a rate of 1010 particles per second coming into the input of the converter we receive approximately 77000 pulses per second at the output of electron multiplier.

To check the efficiency of the converter on the particles of different masses, the input window of the converter has been irradiated by the atoms of hydrogen, helium and argon at the range of energies from 200eV to 2500eV.

Table 1. Values of the ion.

U_0 (keV)	I_0^+ (nA)	$I_{reflected0}$, $\times 10^{10}$ (part/sec)
1000	1500	1.76
700	500	0.87
500	400	0.31
300	150	0.05
200	90	0.04

The measurements of current to a target and pulses counting speed have been conducted at the most efficient operating potentials (300 V on the inner and 500 V on the outer plates), at perpendicular flow of the neutrals into the converter and at the most efficient angle of incidence of 16 degrees to the normal to the plane of the input window of the converter. Hydrogen atoms were obtained in two ways: by neutralization of H^+ ions and neutralization and dissociation over the surface of excimer H^{3+} ions. It should be noted that the energy in the scattered beam of H atoms (in this case) will be 3 times lower. Since the use of different ions changes the intensity of the primary beam on the target, the pulse counting measurements are normalized to the initial current (figure 4). As can be seen, with increasing masses of the incident particles electron multiplier signal intensity (normalized to the initial current) decreases due to the decrease in the overall output of reflected particles entering the converter.

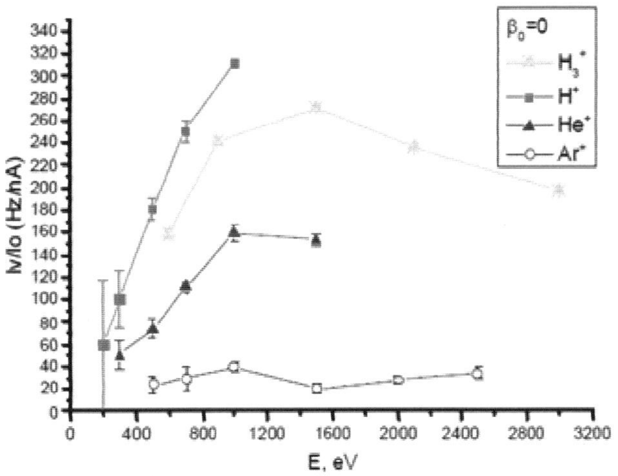

Figure 4. Converter efficiency testing on the particles of different masses.

References

[1] http://www.lr.cosmos.ru/devices
[2] Gott Yu V, Kurnaev V A and Vaisberg O L 2008 *Korpuskularnaya diagnostika laboratornoi plasmy* (Moscow: MEPhI) (in Russian)
[3] Wurz P et al 2006 *e-J. Surf. Sci. Nanotech* **4** 394

Journal of Physics: Conference Series **666** (2016) 012033 doi:10.1088/1742-6596/666/1/012033

Electrode material influence on emission properties of a low inductance vacuum spark

I F Raevskiy[1, 2], **S A Sarantsev** [1]
[1]National Research Nuclear University MEPhI, Kashirskoe sh., 31, Moscow, 115409, Russian Federation

E-Mail: ilya.raevsky@gmail.com

Abstract This paper presents results of a study of the x-ray emission and the composition of the material ejected from the inter-electrode gap of a high-current, low-inductance vacuum spark. Pairs of electrodes made of different materials (Fe, Cu, Mo, W, Pb) were used. The most stable reproducibility of x-ray radiation emission is found to be with the Fe and Mo cathodes.

1. Introduction

A plasma source based on the high-current, low-inductance vacuum spark (LIVS) discharge [1-5] is promising for a large variety of applications. Such plasma sources are well suited for wide spectra of technological processes due to the extreme plasma parameters, the construction simplicity and low cost. Furthermore, studies of the plasma parameters, the emittance characteristics and the detail of the processes in the hot spot (HS) area – highly ionized, high energy, dense plasma formation in the plasma inside LIVS – are of great interest for larger machines due to the similarity of the processes occurring in all z-pinch-based devices.

A large number of factors (such as material and the shape of the electrodes, the stored energy, the discharge current, the initiation method and the trigger energy) affect stability of the plasma parameters, making it hard to optimize LIVS-based devices for a practical application. The lack of full physical understanding of correlation among all the processes in the inter-electrode gap of a LIVS discharge (from the moment of ignition to HS dissipation) makes the systematic work on collecting the data on various factors that can affect LIVS discharge development dynamics extremely important.

X-ray and material emission characteristics of the LIVS discharge plasma with different electrode materials are presented in this paper. Experiments were carried out at NRNU MEPhI Plasma Physics department on the “PION” device [6].

The first stage of commissioning of “PION” included the work aimed at stabilizing HS as the x-ray source. Keeping the interelectrode distance and the single electrode material (iron) fixed, different trigger systems (different energetics, electrode shapes and insulator materials) were tested. The trigger system was found to make a strong impact on reproducibility of the x-ray emission parameters.

Two systems described below – with the “worst” and “best” reproducibility e are shown in figure1. In the “worst” system (b), less than 20% pulses were followed by the x-ray burst, cf. more than 80% in

[2] To whom any correspondence should be addressed.

the "best" one (a). Figure 2 shows the scatter of the starting time and the amplitude of x-ray burst occurrence (the x-ray burst duration is less than 1 ns) from more than 500 discharges. Black and grey areas are for the "best" and "worst" trigger system respectively.

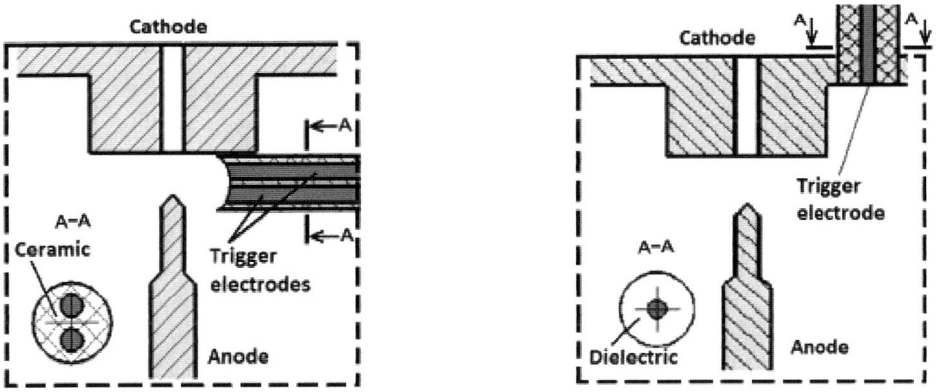

a) b)

Figure 1. Trigger systems schematic: (a) – HS detected in more than 80% pulses, (b) – HS detected in less than 20% pulses.

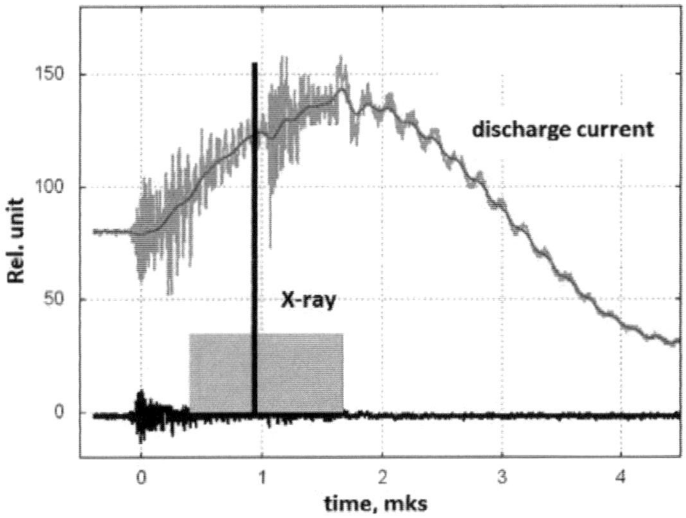

Figure 2. HS x-ray intensity and time of appearance range in series.

The electric parameters of the selected trigger subsystem are the following: the peak voltage is 40 kV, the pulse duration 400 ns, the front rising rate 2×10^{11} V/s and the peak current 100 A. The HS formation moment was detected by the Rogowski coil (in the current transformer mode) and the x-ray pin-diode (hν > 5 keV).

2. Electrode material influence on the x-ray emission characteristics

To determine the electrode material influence on the x-ray emission properties of LIVS, several pairs of the electrodes (anode-cathode) were used: Cu-Pb, Cu-Cu, Cu-Fe, Cu-Mo, Cu-W, Fe-Cu, Fe-Fe, Fe-Mo, Fe-W. The experiments were conducted using the "best" trigger system (figure 1(a)). 100 pulses were shot for each pair. The operating voltage was 15kV.

The Cu anode exhibits the maximum HS appearance probability in the cases with the Fe and Mo cathode (90% and 70% respectively). This pair shows the minimum scatter of the HS occurrence time. No HS were detected in the case of the Pb cathode.

The Fe anode showed similar results for the HS appearance probability (100% and 80% for Fe and Mo cathode, respectively) with the minimum scatter of the HS occurrence time in this study. Table 1 presents the HS formation probability for the material combinations studied.

Table 1. LIVS discharge x-ray emission.

Anode material	Cathode material	Fraction of discharges in the HS mode (%)	Maximum magnitude of the x-ray pulse (a.u.)	X-ray radiation emission timepoint (ns)
Cu	Cu	45	70	1000±300
Cu	Pb	0	-	-
Cu	Fe	90	140	900±100
Cu	Mo	70	100	900±100
Cu	W	40	60	1000±300
Fe	Cu	50	70	900±200
Fe	Fe	100	150	950±50
Fe	Mo	80	130	900±100
Fe	W	40	180	900±150

3. Composition of the material eroded from the electrode and ejected from the inter-electrode gap

Data on the electrode erosion material were investigated. Intense material flow from the inter-electrode gap could be the source of the fast film deposition on the diagnostic windows (quartz or beryllium). This deposition can change the parameters of the input (laser diagnostics) or output (visual, UV, x-ray) radiance. This study yields information about the rate of material deposition on the structures inside the vacuum chamber at different distances from the plasma source.

For this analysis, substrates made of laboratory glass were used. Four substrates were used for each experimental series at four different positions (40, 57, 140 and 210 mm from the gap). These substrates were subjected to 50 pulses. Scanning electron microscopy was then used to analyze their surface.

Scanning electron microscopy was carried out using HITACHI TM-1000 with maximum magnification of 10000. This limitation did not allow us to see granulation of the refractory electrode materials (W, Fe, Mo) because the deposited structures were smaller than the resolution of the microscope, whereas the element analysis detected the presence of the electrode material there. Carbon was always found on the surface, making not more than 7-8% of the deposits (the microscope cannot detect carbon).

The results with the Cu anode for the substrates at the 40 mm distance are shown below. At this distance, strong electrode material deposition was detected for all the electrode pairs. At 210 and 140 mm noticeable deposition was only seen with the Cu and Pb cathodes. At the distance of 57 mm from the inter-electrode gap, there was a strong deposition on the substrates in all cases. But in the cases with the W and Fe cathodes there were glass materials (Si, Ca) on the substrate surface. This means that the deposited layers are more inhomogeneous and have a smaller thickness in these cases.

Figure 3 presents images of the substrate surface (the distance from the interelectrode gap is 40 mm) with the Cu anode and Pb cathode. The image portraits a solid Pb coating on the substrate surface (the element analysis showed 100% Pb). The surface is coated with cellular formations of the size of approximately 600 nm. These formations are apparently droplets of the melted cathode material. Bigger drops (5-10 μm) with the same small structures on them are also seen on the background of the uniform coating.

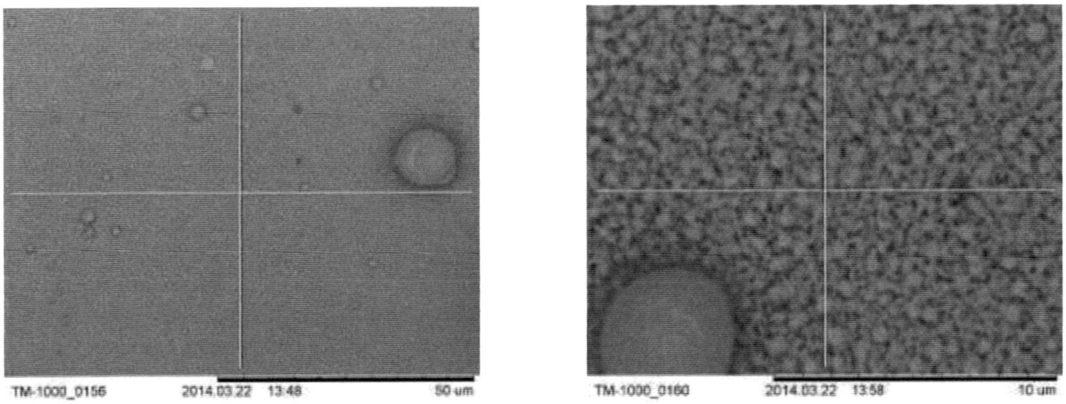

×1800 ×10000

Figure 3. Substrate surface image (40 mm distance) with the Pb cathode.

Table 2. Element compound of the test substrate surface.

Cathode material	Anode material, %(Cu)	Cathode material, %	Silicium, %	Calcium, %	Cell size on surface of test substrate, nm
Pb	0	100	0	0	700±100
Cu	100	100	0	0	500±100
Fe	15÷26	37÷40	22÷27	10÷15	400±100
W	22÷51	0÷49	23÷30	10÷17	<200

Deposition patterns are nearly the same for all the cathode materials tested – the difference is only in the cell size (see table 2) and the presence of the test substrate material on the surface.

4. Conclusions

The selection of the cathode material strongly affects the x-ray emission characteristics. The best x-ray reproducibility from pulse to pulse is registered for the Fe and Mo cathodes. A stable pinch effect with a relatively small time scatter (100-200 ns) is detected for these cathodes.

The analysis of the corpuscular flow composition (from scanning electron microscopy of the glass substrates) also reveals a strong dependence on the cathode material selection. The predominantly cathode material is detected in the case of the low melting temperature of the cathode material (Pb). With higher melting temperature (Fe and Mo) the percentage of the cathode material decreases, whereas the fraction of the anode material (Cu) grows. In the case of the Fe or W cathode there are glass composition elements on the substrate surfaces. This indicates a lower thickness of the layers deposited on the substrate surface.

References

[1] Kuznetsov A P, Byalkovskii O A, Gubskii K L, Kozin G I, Protsenko E D, Dodulad E I and Savjolov A S 2014 *Plasma Physics Reports* **40** 290-297
[2] Beilis I I, Boxman R L, Goldsmith S and Paperny V L 1999 *Applied physics letters* **75** 2734
[3] Krasov V I, Paperny V L , Korobkin Yu V, Romanov I V, Rupasov A A and Shikanov A S 2007 *Technical physics letters* **33** 941
[4] Bashutin O A, Alkhimova M A, Vovchenko E D, Dodulad E I, Savelov A S and Sarantsev S A 2013 *Plasma Physics Reports* **39** 900-909
[5] Baronova E O et al. 2012 *Plasma Physics Reports* **38** 751-760
[6] Sarantsev S A, Dvoyeglazov Ya M and Raevskiy I F 2015 *Physics Procedia* **71** 133-137

Testing composite parametrical hypotheses without applying the reduction

V Vlasov[1], S Vlasova[2] and A Tolokonsky[1,3]
[1] National Research Nuclear University MEPhI (MEPhI), Kashirskoe shosse 31, Moscow, 115409, Russian Federation
[2] All-Russian Research Institute for Nuclear Power Plants, Ferganskaya st., 25, Moscow, 109507, Russian Federation

E-mail: AOTolokonskij@mephi.ru

Abstract. Usually when the parametrical hypotheses are being tested the Wald reduction from composite hypothesis to the simple one is used. However, in order to apply the reduction method it is needed to know the distribution law of unknown parameter. Practically such a law cannot be determined precisely using the experimental data. First of all, it requires long supervision over the controlled stationary process to provide an inalterability of probability characteristics of the process. In practice the modifications in technological process could be made, and therefore the probability characteristics of the process can also change. Using the example of exponential distribution the algorithm of testing composite parametrical hypothesis about the distribution parameter which does not exceed the declared threshold value without use of reduction is considered in article. Such approach is based on the fact that the partition boundary of the sample space depends monotonously on unknown value of the interest parameter.

1. Introduction

Usually when the parametrical hypotheses are being tested the Wald reduction from composite hypothesis to the simple one is used. However, in order to apply the reduction method it is needed to know the distribution law of unknown parameter. Practically such a law cannot be determined precisely using the experimental data. First of all, it requires long supervision over the controlled stationary process to provide an inalterability of probability characteristics of the process. In practice the modifications in technological process could be made, and therefore the probability characteristics of the process can also change.

2. Experimental setup

Let us assume that the threshold value of parameter λ is λ_0. On the basis of the sample $\xi_1, \xi_2, ..., \xi_n$ of the population which is described by the following probability density function:

$p(x) = \begin{cases} \lambda \exp(-\lambda x), x \geq 0 \\ 0, x < 0 \end{cases}$, it is needed to test the statistical hypothesis H_1, which claims the inequality $\lambda < \lambda_0$ is true (this inequality defines the composite parametrical hypothesis H_1). In order

[3] To whom any correspondence should be addressed.

to solve the task we will consider the alternative composite parametrical hypothesis H_2, that states the other inequality $\lambda > \lambda_0$ is true.

Such a task can appear in a wide range of application areas. For example, in the analysis of scientific experiments or in the analysis of unsuccessful changes in technological processes resulting in deterioration of production etc.

At first we will consider the order of constructing the critical region basing on Pearson's theorem in order to test two simple competing hypotheses H_1, H_2. The critical region for a hypothesis H_1 (when this hypothesis is rejected) is defined in the sample space by the inequality:

$$p(x_1, x_2, ..., x_n, H_2) > Cp(x_1, x_2, ..., x_n, H_1), \quad (1)$$

where: $p(x_1, x_2, ..., x_n, H_2)$ - the distribution density of a sample set in the conditions of the hypothesis H_2, $p(x_1, x_2, ..., x_n, H_1)$ - the distribution density of a sample set in the conditions of the hypothesis H_1. A value of a constant C could be chosen both according to the values of probabilities of type I and II errors [1,2], or according to the values of weights (costs) of type I and II errors on the basis of variation principles [3].

It is important to note that when the decision between two parametrical hypotheses H_1 (when $\lambda = \lambda_1$) and H_2 (when $\lambda = \lambda_2$) is being made the values λ_1 and λ_2 should be known. In the considering case the hypothesis H_1 is completely defined: $\lambda = \lambda_0$. In the conditions of the opposite hypothesis H_2, that states the truthfulness of inequality $\lambda > \lambda_0$, the value of parameter λ remains unknown. Therefore the simple parametrical hypothesis competes with the composite one.

We will further use the property of monotonous dependence of the critical region's boundary on the value of unknown parameter λ in the conditions of the composite parametrical hypothesis H_2 which include the set of simple hypotheses. The property of monotonous dependence is considered in the work [4].

Let us assume that in the conditions of the exponential distribution when the hypothesis H_1 is true the value of parameter is known: $\lambda_1 = \lambda_0$. Therefore for any value of the parameter λ in the conditions of the hypothesis H_2 the expression (1) can be formulated in the following way:

$$\lambda^n \exp(-\lambda S) > C\lambda_0^n \exp(-\lambda_0 S), \quad (2)$$

where $S = \sum_{i=1}^{n} x_i$.

The hypothesis H_1 (when $\lambda = \lambda_0$) will be rejected if the inequality (2) is true. If for any value of parameter λ that exceeds the value λ_0 the opposite inequality is true:

$$\lambda^n \exp(-\lambda S) < C\lambda_0^n \exp(-\lambda_0 S), \quad (3)$$

then any simple hypothesis that is included in composite hypothesis H_2 will be rejected. Therefore the composite hypothesis H_2 should be rejected and we come to the following conclusion: when the last inequality is true the value of unknown parameter λ does not exceed the value λ_0. Further we will consider the simplest case when $C = 1$. In this case the inequality (3) can be expressed by the formula:

$$S > \frac{n \ln \gamma}{\lambda_0(\gamma - 1)}, \text{ where } \gamma = \frac{\lambda}{\lambda_0}.$$

Using the concept of sufficient statistics for exponential distribution: $S = \sum_{i=1}^{n} \xi_i$, where ξ_i – sample values, the received formula can be written in a more compact way.

The boundary value S_g for the variable S, which splits the sample space into two parts, could be found from the expression (2) when the equation is reached:

$$S_g = \frac{n}{\lambda_0} \frac{\ln\gamma}{\gamma - 1}, \tag{4}$$

where $\gamma = \frac{\lambda}{\lambda_0}$, $\lambda > \lambda_0$.

Therefore the criterion according to which it is necessary to reject any simple parametrical hypothesis when $\lambda > \lambda_0$, could be simply expressed by the inequality:

$$S > S_g. \tag{5}$$

This directly means that the parameter λ does not exceed the value λ_0. The simple analysis shows that the variable S_g decreases when the parameter γ in its turn increases. Therefore if for some value γ^* the criterion (5) is true, it is also be true for every values of γ, for which $\gamma > \gamma^*$. If $C = 1$ and γ tends to the value 1 the expression (4) is uncertain. This problem can be easily overcome by means of application of the wide known Lopital rule [6], and the variable S_g could be found from the expression: $S_g^* = \frac{n}{\lambda_0}$. Actually, when $\gamma \to 1$ the limit will be equal to the expression:

$$\lim \frac{n}{\lambda_0} \frac{\ln\gamma}{\gamma - 1} = \frac{n}{\lambda_0} \lim \frac{\frac{1}{\gamma}}{1} = \frac{n}{\lambda_0}.$$

Therefore the criterion that the parameter λ does not exceed the value λ_0 in this case has a simple expression:

$$S \le \frac{n}{\lambda_0}. \tag{6}$$

The condition $C = 1$ corresponds to the case when the weights (costs) of type I and II errors are equal [7].

The dependence of variable S_g on parameter λ for the case when $C = 1$ is presented in figure 1.

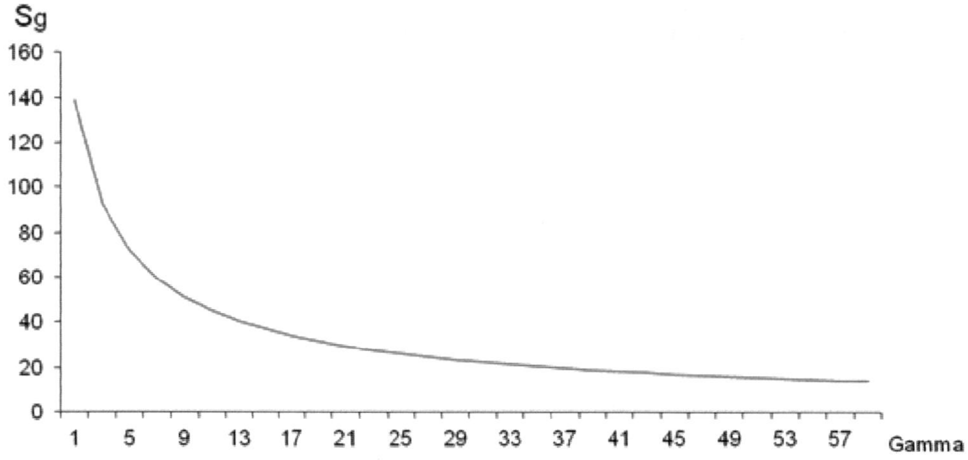

Figure 1. The dependence of variable S_g on parameter λ for the case when $C = 1$.

The situation when $S > \frac{n}{\lambda_0}$ also needs detailed consideration. In this case it is necessary to consider the parametrical hypotheses which claims that $\lambda < \lambda_0$. In the work [4] it was shown that when λ decreases (that is when $\lambda < \lambda_0$) the variable S_g increases, and if the parameter λ tends to zero the variable S_g tends monotonously to infinity (in this work the following mistake was made: the incorrect sign < was used instead of correct sign > when the hypotheses were tested, and therefore this led to the opposite conclusion). When the values of the parameter λ are small this hypothesis should be rejected (when the condition $S < S_g$ is true, so it cannot be accepted). It could be simply explained by the following illustration: let us assume that the value of the parameter λ is very small, then the sample values have to be very big and the hypothesis claiming that the values of parameter λ are small will be true if the sum S of sampled values is rather great. We will show it mathematically. Let us denote H as the hypothesis that states $\lambda = \lambda_1$, where $\lambda_1 < \lambda_0$, and by symbol $\overline{H}$ the hypothesis that states $\lambda = \lambda_0$.

According to condition (1) the hypothesis H is rejected in the case when the following inequality is true:

$$\lambda_0^n \exp-(\lambda_0 S) > C\lambda_1^n \exp-(\lambda_1 S).$$

Applying simple algebraic transformations to the last expression lead to the result:

$$S < \frac{n \ln \frac{\lambda_0}{\lambda_1} - \ln C}{\lambda_0 - \lambda_1} = S_g$$

The main result about the monotonous dependence of the critical region's boundary on the value of the parameter λ and possibility of its applying for testing composite parametrical hypotheses about the value of the parameter λ remain true. Therefore any simple parametrical hypothesis should be rejected when $\lambda = \lambda_1$ is quite small in case of truthfulness of the inequality $S < S_g$. It is clear, that if the hypothesis λ^* is rejected, then any of the simple hypotheses claiming that $\lambda < \lambda^*$ should be also rejected. Obviously the acceptance of the hypothesis about small values of the parameter λ requires large sample size and considerable time for supervision.

The dependence of the boundary S_g on the parameter λ in case when $C = 1$ and the value of the parameter λ is quite small is shown in figure 2.

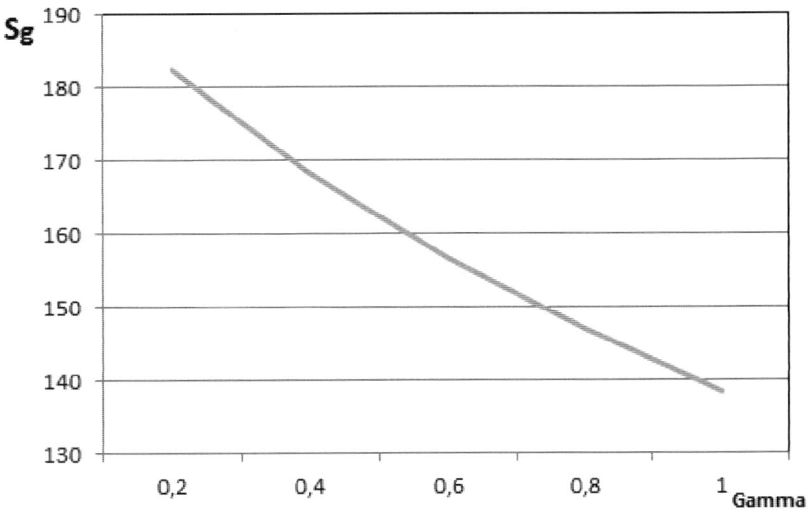

Figure 2. The dependence of the boundary S_g on the parameter λ in case when $C = 1$ and the value of the parameter λ is quite small.

3. Discussion

Therefore the whole sample space is split into two regions and in every region the certain hypotheses are accepted. It should be noted that notwithstanding the acceptance of any of composite hypotheses the true value of the parameter λ remains unknown. For research purposes in order to avoid difficulties of the analysis when C is not equal to 1, it is possible to consider the auxiliary parametrical hypothesis which claims that the value of parameter λ is equal to the value λ_0', which is close to λ_0. This question deserves further more detailed discussion.

References

[1] Wilks S S 1961 A combinatorical tests for the problem of two samples from continuous distribution *Proc. Fourth Berkeley Symp. On Math. Stat. and Prob.* vol 1 (University of California Press.)
[2] Nikitin A I, Leipunsky I O and Nikitina T F 2010 *Prikladnaya Fizika* **2** 15-22 (in Russian)
[3] Rao C R 1968 *Linear Statistical Methods and its Applications* (M: Science) (in Russian)
[4] Vlasov V A 2012 *Estimates, Decisions, Risks* (M., Binom) (in Russian)
[5] Vlasov V A, Vlasova S V and Tolokonsky A O 2014 Estimates when parametrical hypotheses are accepted and comparison them with maximum likelihood estimates *Life Science Journal* **11**.
[6] Seber G A F 1977. *Linear Regression Analysis, Wiley Series in Probability and Mathematical Statistics* (New York-London-Sydney-Toronto) pp:456
[7] Fikhtengol'ts G M 2003 *Course of Differential and Integral Calculus* (M., Fizmathlit) (in Russian).
[8] Vlasov V A and Vlasova S V 2007 Software Support for Testing Composite Hypotheses about the Parameters of Exponential Distribution taking into Account the Weighs of Errors *Instruments and Systems: Monitoring, Control, and Diagnostics* **8** 25-27 (in Russian)

Diagnostics of beam plasma produced in dielectric cavity at fore-vacuum pressures

D. Zolotukhin[1]
Tomsk State University of Control Systems and Radioelectronics, Lenina prospekt, 40, Tomsk, 634050, Russia

E-mail: ZolotukhinDen@gmail.com

Abstract. The results of the probe and optical measurements of plasma parameters produced during the injection of the electron beam with the energy of 3-8 keV and current of 30-60 mA into a dielectric cavity at the argon pressure of 2-12 Pa are presented. The possibility of continuous injection of the electron beam into the cavity without current-collecting electrodes inside is demonstrated. The axial distribution of the values of the plasma parameters (density, electron temperature and floating potential) in the cavity is discussed. The plasma concentration in the cavity is shown to be approximately 20% higher, and the electron temperature 1.5-2 times higher, than in the case of generating the plasma in the free space of the vacuum chamber.

1. Introduction

Creating plasma in the cavity bounded by the dielectric walls is important for ion-plasma modification of the internal surfaces of glass or plastic vessels [1]. Electrodeless HF or microwave discharge in a magnetic field is typically used to do this [2]. The disadvantage of this approach is the low efficiency of energy transmission from the source to the HF plasma and the limited range of operating pressures. Electron beam plasma generation is largely free of these shortcomings. However, injection of the beam into the cavity with pressure parameters traditional for the electron sources (10^{-2} Pa) is difficult because of the negative charge accumulation inside the cavity, and the use of a differential pumping system greatly complicates the device. Application of the fore-vacuum plasma electron sources [3] operating at pressures of 1-100 Pa solves the problem of charge accumulation in the cavity owing to formation of a dense plasma in the electron beam transport area. The ions of this plasma neutralize the charge of the accelerated electrons accumulated on the dielectric surface [4]. The purpose of this study was to measure the parameters of the plasma generated in the fore-vacuum pressure range in a dielectric cavity during injection of the electron beam into the cavity.

---

[1] To whom any correspondence should be addressed

Journal of Physics: Conference Series **666** (2016) 012035 doi:10.1088/1742-6596/666/1/012035

2. Experimental setup

Experiment scheme is shown in figure 1.

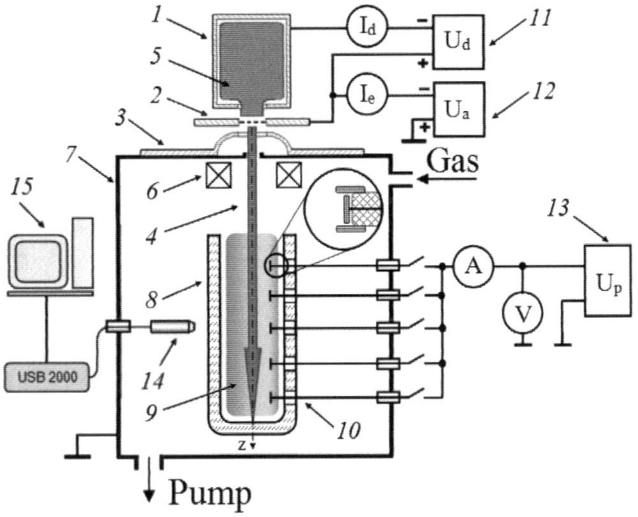

Figure 1. Experiment schematic: *1* – hollow cathode, *2* – anode, *3* – accelerating electrode (extractor), *4* – electron beam, *5* – emission plasma, *6* – focusing system, *7* – vacuum chamber, *8* – quartz cavity, *9* – beam plasma, *10* – Langmuir probe, *11* – discharge power source, *12* – accelerating voltage source, *13* – probe bias voltage source, *14* – light guide of the USB 2000 spectrometer, *15* – personal computer.

The electron beam (4) was created by a fore-vacuum plasma electron source based on a glow discharge with a hollow cathode [3] in the CW mode. The discharge in the source was maintained by the voltage U_d = 300 - 500 V between the hollow cathode (1) and the anode (2) with the discharge current I_d in the range of 100 to 400 mA. The electron beam was extracted from the discharge plasma (5) through the holes in the anode (2), accelerated to energies of 3 - 8 keV by the voltage U_a in the gap between the anode and the grounded extractor (2-3), and then focused by the magnetic field of the coil (6). The beam current was estimated from the current I_e in the circuit of the accelerating voltage source (12) and was 30 - 60 mA. The beam diameter was 4-6 mm. Vacuum in the chamber (7) was sustained by the mechanical fore-vacuum pump ISP-1000C, pressure of the working gas – argon or air - was set to 2-12 Pa and regulated by the inlet valve. Accelerated electron beam was injected into the cylindrical quartz cavity (8) with the inner diameter of 40 mm, length of 200 mm and wall thickness of 2 mm, and created the plasma (9) inside the cavity. To study the distribution of the beam plasma parameters along the z-axis, five identical single, plane Langmuir probes (10) were inserted into the cavity through the 6 mm perforations made in the side wall. The diameter of the receiving surface of each probe was 3 mm. The distance from the axis of the beam to the receiving surface of the probe was 15 mm. In order to prevent the contact of the fast electrons of the beam with the probes, each probe was surrounded with a protective metal cylindrical screen having the diameter and length of 5 mm and being under the floating potential. Set of probes was firmly fixed with respect to the axis of the beam and the walls of the chamber, and the cavity could be extracted to identify differences in the behavior of the plasma inside the cavity and in conditions of free beam transport inside the chamber. Plasma emission was studied using the optical spectrometer USB-2000 from Ocean Optics. The light guide (14) of the spectrometer was located approximately in the middle of the cavity at a distance of 1.5 cm from its open end. Spectrometer signal was processed by a specialised software on the computer (15).

Beam plasma parameters were determined from the probe characteristic. The electron temperature T_e was determined on the exponentially growing e-branch in the area between the floating potential of the probe and the plasma potential. The value of the probe voltage corresponding to the zero second derivative of the probe current with respect to the voltage was taken as the plasma potential. The plasma concentration n_i was determined by the saturation current on the ion branch of the probe characteristic.

3. Results and discussion

The results of the measurements of the floating potential of the probe are shown in figure 2.

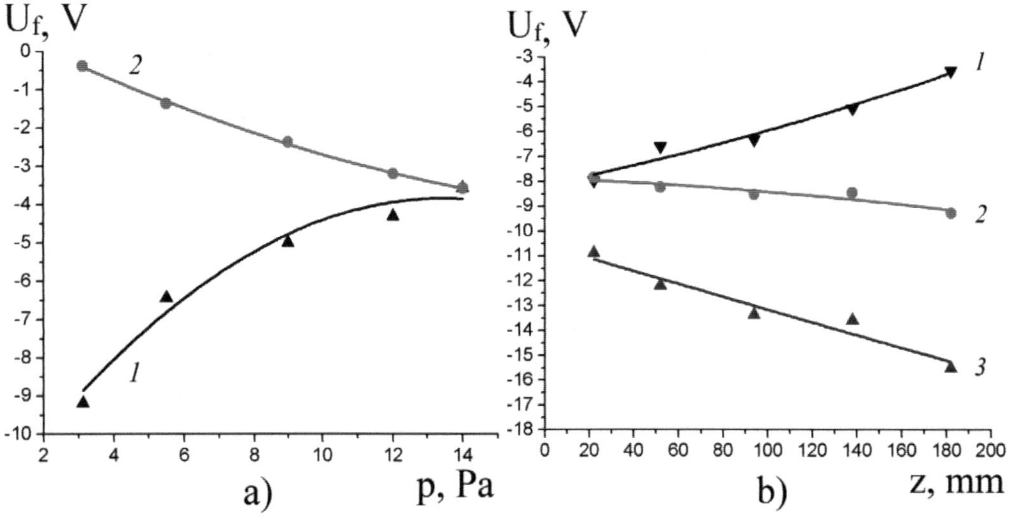

Figure 2. a) Floating potential of the probe U_f(z = 52 mm) in the cavity (*1*) and in free space (*2*) vs. pressure,; b) axial distribution of the floating potential in the cavity for the pressure values of 12 Pa (1), 6 Pa (2) and 2.5 Pa (3). Experimental conditions: I_e = 60 mA, U_a = 3 kV; working gas: argon. Zero of the z-axis corresponds to the open end of the cavity.

In the case when the plasma is confined in the cavity (Figure 2a, curve 1), the U_f is lower and decreases along with the reduction of the pressure. This might be explained by accumulation of the electrons inside the cavity because the conditions of losing the charge deteriorate progressively as the plasma density decreases. The U_f increase with the pressure reduction in the absence of the cavity (figure 2a, curve 2) can be associated with a more efficient removal of the electrons from the plasma region due to the reduction of the electron-atom collisions. Character of the axial distribution of the floating potential in the cavity, as shown in figure 2b, depends on the pressure. For example, at 12 Pa, U_f value weakly grows inward the cavity, and at 6 and 2.5 Pa this value decreases. Such a behavior indicates a change of the axial electric field distribution in the plasma by the pressure change and requires further study. Comparison of the axial distributions of concentration and temperature of the plasma electrons created in the cavity with those in the plasma produced in the absence of the cavity is shown in figure 3. At the beam energy of 3 keV, the plasma density in both cases decreases along the beam direction, but is approximately 20% higher in the cavity (figure 3a, curve 1). This could be attributed to additional ionization of the gas inside the cavity by the secondary electrons which are dislodged from the surface by the beam electrons and the plasma ions and accelerated in the Debye sheath.

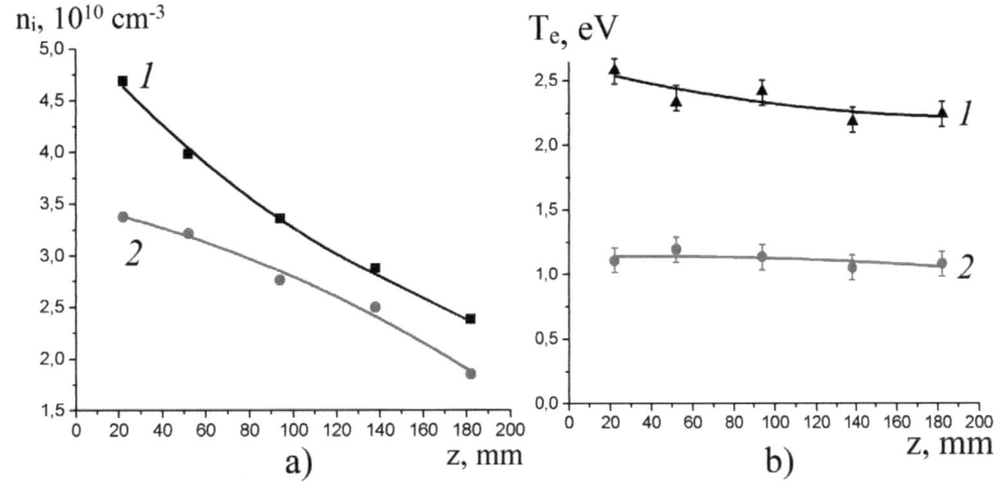

Figure 3. Axial distributions of the electron density (a) and temperature (b) for the plasma created inside the cavity (*1*) and in the free space of the chamber (*2*). Pressure is 6 Pa, I_e = 40 mA, U_a = 3 kV.

The electron temperature in both cases is nearly independent of z, but the electron temperature is 1.5 - 2 times higher in the cavity. This can be explained by a reduction of the outflow of the plasma energy from the volume of plasma in the cavity compared to the case of the absence of the cavity, since the energy leaves the volume with the plasma particles only through the open end of the cavity. The higher temperature of the plasma electrons in the cavity can be indirectly confirmed by the emission spectra, which are shown in figure 4.

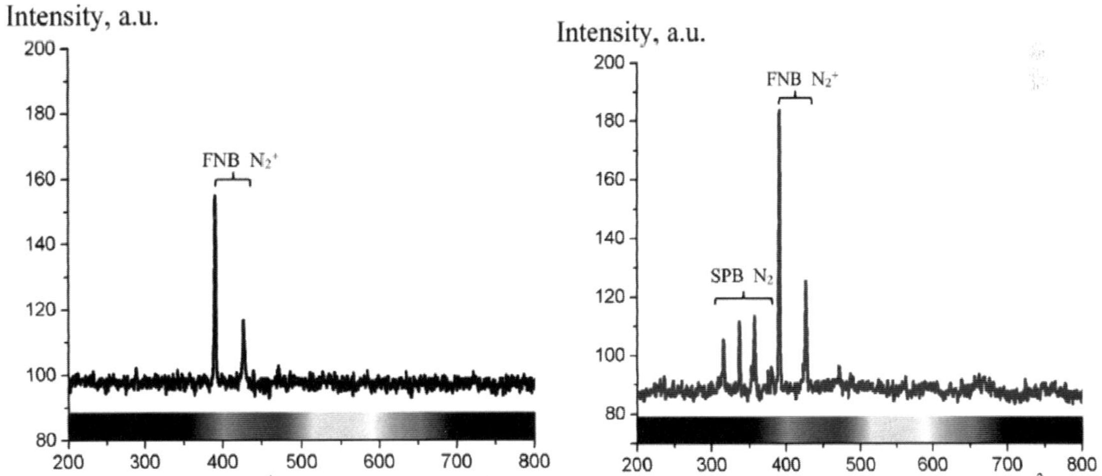

Figure 4. Optical spectra of the beam plasma created in the free space (a) and inside the quartz cavity (b). Pressure is 3.5 Pa, beam current I_e = 30 mA, U_a = 8 kV; working gas: air.

As follows from figure 4a, in the case of beam plasma creation in the chamber without cavity, the 391 and 428 nm lines appear in the plasma emission spectrum. These lines are specific of the First Negative Band (FNB) of molecular nitrogen ions N_2^+, the excitation potential of which is rather high, about 20 eV [5]. Excitation of this system can attributed to the energetic electrons of the beam. At the same parameters of the experiment, in the case of plasma generation in the dielectric cavity, as shown in figure 4b, the 315, 336 and 357 nm lines appear in the emission spectrum in addition. These lines belong to the Second Positive Band (SPB) of the excited nitrogen molecules N_2 with a lower excitation potential of about 13 eV [6]. By increasing the gas pressure to 12.5 Pa, the SPB lines appear also in the emission spectrum of the "open" beam plasma. The observed features of the optical spectra can be explained as follows. The generation rate for the nitrogen molecules can be described by the following formula:

$$\frac{dn^*}{dt} = n_e n_g K(T_e) \quad (1)$$

where n_e, n_g are the densities of the plasma electrons and gas molecules, respectively, and $K(T_e)$ the reaction rate constant that is directly proportional to the electron temperature in the beam plasma.

Since T_e is lower in the "open" beam plasma than in the quartz cavity (figure 3b), even with n_e and n_g held approximately equal, the rate of generation of molecular nitrogen ions with the excited SPB will be much lower than in the quartz cavity due to the lower $K(T_e)$. It is known that the electron temperature generally decreases along with the pressure increase and the plasma concentration increases, so the number of the molecular ions with excited SPB increases due to the increase of the plasma density with pressure.

4. Conclusion

The results of the probe and optical measurements illustrate the difference in the characteristics of the behavior of the beam-generated plasma (in terms of the floating potential, density and temperature) in the cavity compared and that in the free space. The parameters of the plasma are suitable for using it in technological applications.

5. Acknowledgements

The work was supported by RFBR, grant №16-38-00230 mol_a. The author wishes to thank M.I. Lomaev (IHCE SB RAS) for his help with the optical spectra analysis.

References

[1] Sakudo N, Ikenaga N, Ikeda F, Nakayama Y, Kishi Y, Yajima Z, Jiro Matsuo, Masataka Kase, Takaaki Aoki and Toshio Seki 2011 *AIP Conference Proceedings* ***1321*** 266
[2] Conrads H and Schmidt M 2000 *Plasma Sources Sci. Technol.* **9** 441
[3] Burdovitsin V A and Oks E M 2008 *Laser and particle beams* **26** 419
[4] Burdovitsin V A, Klimov A S and Oks E M 2009, *Technical Physics Letters* **35** 511
[5] Stewart D.T. 1956 *Proc. Phys. Soc. A* **69** 437
[6] Thompson N and Williams S E 1934 *Proc. Roy. Soc. A* **147** 583

Laser plasma influence on the space-time structure of powerful laser radiation

O B Ananyin[1], G S Bogdanov[1], E D Vovchenko[1,2], I A Gerasimov[1], A P Kuznetsov[1] and A P Melekhov[1]
[1] National Research Nuclear University MEPhI (MEPhI), Kashirskoe shosse 31, Moscow, 115409 Russian Federation

E-mail: edvovchenko@mail.ru

Abstract. This paper deals with the influence of laser plasma on the structure of the radiation field of a powerful Nd-glass laser with pulse energy up to 30 J and with the diameter of the output beam 45 mm. Laser plasma is generated by focusing the laser radiation on a low-density target such as nylon mesh and teflon or mylar films. Temporal profile of the laser pulse with a total duration of 25 ns consists of a several short pulse train. Duration of each pulse is about 2 ns. Notable smoothing of spatially non-uniform radiation structure was observed in the middle of the laser pulse.

1. Introduction

Laser fusion studies face a number of problems. One of them is elimination of instabilities, which are formed during the compression of a thermonuclear target for its non-uniform irradiation with laser light. This heterogeneity is typically caused by the laser pulse generation process and remains in its further amplification and transmission through the channels of a powerful laser system. Currently, to smooth the profile of the laser beam intensity the passage of a laser beam through a phase plate is widely used. These plates are an optically inhomogeneous medium with a refractive index that varies quite rapidly in space and time. A desired structure of the medium in the form of dynamic plasma phase plate (DPPP) can be obtained based on a laser plasma, which is formed on a low-density target in the area of focus of the laser beam [1].

During the experiments on a powerful laser facility [2] the DPPP in a form of laser plasma was created with using a preliminary pulse (~50 J, ~1 ns). Diagnostic laser pulse (~15 J, ~1 ns), for which the smoothing was conducted, passed through the plasma with a delay Δt. The maximum smoothing effect was observed at a delay of $\Delta t \sim 150$ ps. Furthermore, the smoothing efficiency has been changing significantly during the laser pulse. The modulation of the laser radiation intensity was significantly smoothed only in the time interval corresponding to the central area of the laser pulse. In this paper, we consider the smoothing feature of the spatial structure of powerful laser radiation, when it consists of several short (~2 ns) pulses.

2. Experimental installation

The studies were conducted on an experimental installation "MISHEN" ("Target"), the scheme for which is shown in figure 1. During the experiments the initial laser radiation characteristics were

[2] To whom any correspondence should be addressed

recorded, as well as the radiation transmitted through the DPPP target: energy, duration and the shape of the laser pulse, the spatial structure of a cross-section of the beam.

Laser system consists of the Q-switched pulse generator (1) on the active element of Nd: phosphate glass (GLS–22) and two amplifier modules (3 and 4) based on the solid-state laser GUK-4507 (GLS–21). Module (3) is made under the scheme of seven-pass telescopic amplifier with phototropic shutter LiF. Module (4) is single-pass. The output of the laser beam (~45 mm) had a pulse energy up to 30 J for the wavelength $\lambda = 1.055$ nm and total duration of the laser train ~25 ns, the power density on the DPPP target was $I \sim 10^{16}$ W/m^2. For visualization of the spatial structure of a cross-section of the laser beam we used a partial conversion (about 1%) of infrared radiation into the second harmonic using the DKDP crystal (5).

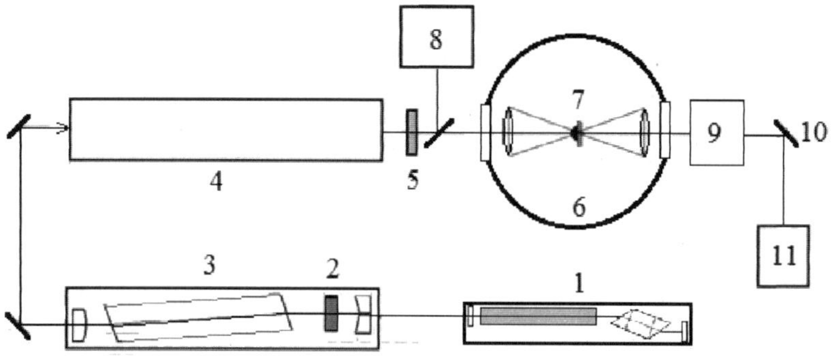

Figure 1. Experimental laser installation "MISHEN" ("Target").

Laser plasma was produced in the vacuum chamber (6) at a pressure of 10^{-3} Pa, and a laser beam was focused onto the DPPP target (7). The focal lengths of the input and output lenses were equal to 120 mm. To control the energy and temporal characteristics of the laser pulse we used standard measuring tools, combined into a diagnostic module (8). The laser radiation that has passed through the laser target, is gathered back into a parallel beam (~ 40 mm) and attenuated into the optic module (9). Next, after a strong attenuation ($I_{in}/I_{out} = 1.5 \cdot 10^4$) the beam is further compressed by 3 times to match the height of the entrance slit of a digital camera «BIFO» K-008 (11). To suppress the infrared radiation the interference mirror (10) was used with a maximum reflection coefficient at the wavelength $\lambda \sim 530$ nm. The second harmonic radiation was recorded with using the K-008 camera, operating in the linear sweep mode with a coefficient of 30 ns/cm («streak» mode) and the height of the slit is 20 mm.

3. Experiment and results

Figure 2 shows the temporal profiles of the initial radiation recorded by a variety of diagnostic tools: 1 – laser pulse (digital streak frame camera K-008, BIFO); 2 – laser pulse (*pin*-photodiode FDUK); 3 – X-ray pulse (*pin*-photodiode, 9 micron Al foil). Comparison of the *pin*-diode signals has shown that the laser plasma emits the X-rays during the entire duration of the laser pulse, about 20 ÷ 25 ns. A more detailed study of the laser pulse was conducted using the high-speed digital camera K-008 having a high time resolution (≤ 1 ns). In this experiment, it was found that the laser pulse is a train of individual short (~2 ns) pulses with intervals between the individual pulses of ~ 4 ns (figure 3).

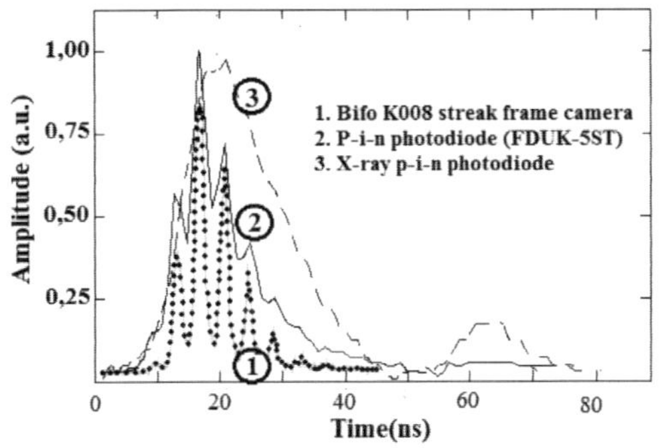

Figure 2. Laser and X-ray pulses.

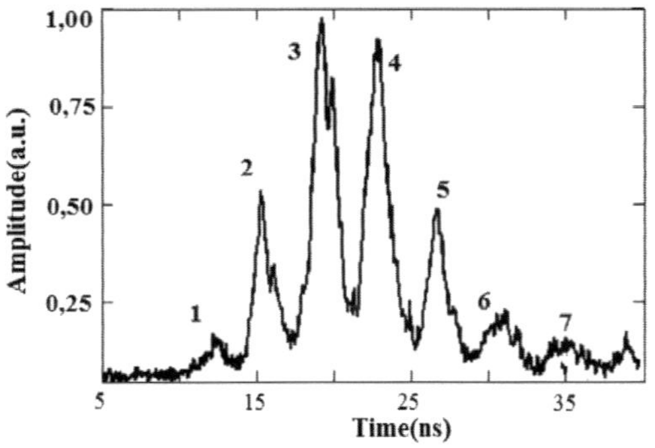

Figure 3. Typical view of laser pulse train.

In a series of comparative experiments using the K-008 camera, an initial space-time structure of the laser radiation was studied along with its variation as a result of passing through the laser plasma. To produce a plasma we used the DPPP targets of different types: 30 micron nylon mesh ($\rho \sim 0.65$ g/cm^3), 15 micron films of teflon ($\rho \sim 2.2$ g/cm^3), and 3 micron film of mylar ($\rho \sim 1.4$ g/cm^3). Intensity profiles for different time moments within the laser pulse train were restored using the initial data.

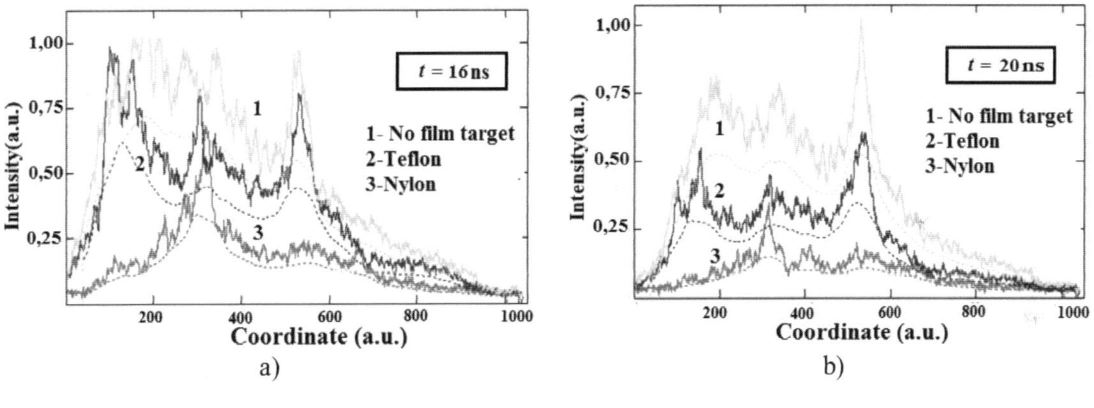

Figure 4. Intensity profiles for the teflon film and nylon mesh targets: a) t = 16 ns; b) t = 20 ns (solid line – experiment; dashed line – estimated smoothing).

As an illustration, figure 4 shows the laser radiation intensity profiles for the three experimental cases (without plasma, plasma on a teflon film, and plasma on a nylon mesh) for instants t = 16 ns and t = 20 ns, which correspond to the second and third pulses in figure 3. The spatial coordinate in the beam cross-section is presented in the relative units (1000 relative units correspond to the diameter of the laser beam, which is 40 mm). Attenuation of radiation intensity of the laser beam passing through the plasma is clearly seen. The experimental data (solid lines) is smoothed by mathematical methods (dashed lines).

To analyze the effect of laser plasma on the space-time structure of the radiation the following processing algorithm was used. The area under the curve corresponding to experimental profile of intensity was calculated along with the area under the curve corresponding to mathematical procedure of smoothing (figure 4). Their difference allows to assess the level of speckle modulation of the laser beam, which determines the noise component in the intensity profiles. Applying the same procedure throughout the duration of the laser train provides an integral noise characteristic.

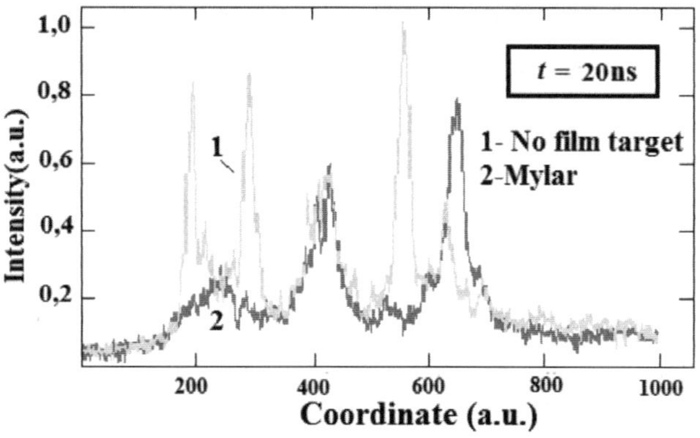

Figure 5. Intensity profiles of the initial radiation (without plasma) and radiation passing the laser plasma obtained with mylar film.

Comparison of the results obtained by this method shows that the integral noise characteristics for the selected laser targets are not very different from similar characteristics of the initial radiation. This indicates the absence or bad quality of smoothing across the entire train of laser pulses. However, in the case of a thin film of mylar (3 microns) for some time moments and some areas of space profile we can see quite good suppression of the speckle modulation (figure 5). Moreover, the smoothing effect is getting stronger in the middle of the laser train.

4. Conclusion

The technique was developed and the preliminary results were obtained for the effect of laser plasma on the spatial and temporal characteristics of high-power laser radiation. Thus for the moment it is not clear how the material, thickness and density of the irradiated sample affect the ability of plasma to locally weaken the separate elements of the speckle pattern. The mechanism of attenuation of the individual speckle is obviously connected with the local plasma parameters. It would be helpful to combine the use of the streak camera with the use of the X-ray diagnostic and laser interferometry.

References

[1] Bessarab A V et al. 1996 *Proc. SPIE* **2767** 97
[2] Voronich I N, Garanin S G, Derkach V N, Zaretskii A I, Kravchenko A G, Lebedev V A, Pinegin A V, Sosipatrov A V and Sukharev S A 2001 *Quantum Electronics* **31** 970

Institute of Physics
Dirac House, Temple Back
Bristol BS1 6BE UK

ISSN: 1742-6588
ISBN 978-1-5108-1958-0

CURRAN ASSOCIATES INC.
proceedings.com